João Paulo Dantas de Carvalho

Hydrolysis of cashew bagasse to obtain bioethanol

João Paulo Dantas de Carvalho

Hydrolysis of cashew bagasse to obtain bioethanol

An experimental analysis and application of the Michaelis-Menten equation

Imprint
Any brand names and product names mentioned in this book are subject to trademark, brand or patent protection and are trademarks or registered trademarks of their respective holders. The use of brand names, product names, common names, trade names, product descriptions etc. even without a particular marking in this work is in no way to be construed to mean that such names may be regarded as unrestricted in respect of trademark and brand protection legislation and could thus be used by anyone.

Cover image: www.ingimage.com

This book is a translation from the original published under ISBN 978-620-6-75842-6.

Publisher:
Sciencia Scripts
is a trademark of
Dodo Books Indian Ocean Ltd. and OmniScriptum S.R.L publishing group

120 High Road, East Finchley, London, N2 9ED, United Kingdom
Str. Armeneasca 28/1, office 1, Chisinau MD-2012, Republic of Moldova, Europe
Printed at: see last page
ISBN: 978-620-7-27670-7

EXPERIMENTAL ANALYSIS AND APPLICATION OF THE MICHAELIS-MENTEN EQUATION TO THE HYDROLYSIS KINETICS OF CASHEW BAGASSE TO OBTAIN BIOETHANOL

Dr. João Paulo Dantas de Carvalho

SUMMARY

HYDROLYSIS KINETICS OF CASHEW BAGASSE TO OBTAIN BIOETHANOL

Recently, the global demand for fuel ethanol has been expanding very rapidly, and almost all fuel ethanol is produced by fermenting sucrose in Brazil or corn glucose in the United States, but these raw materials will not be enough to satisfy international demand. Agro-industrial lignocellulosic residues, such as cashew bagasse, sugarcane bagasse, soybean hulls, etc., are abundant and low-cost sources for the biotechnological production of compounds with high added value, such as ethanol. In this context, the aim of this work was to evaluate the production of bioethanol from cashew stalk bagasse. To this end, the material was initially characterized in order to assess its cellulose and lignin content. After this stage, the activity of the enzyme used was evaluated to confirm the manufacturer's data. Subsequently, ethanol production was studied using lignocellulosic material firstly, the pre-treatment of cashew bagasse with sulphuric acid and phosphoric acid, both diluted, was studied, evaluating reaction time, acid concentration and enzymatic activity in the hydrolysis, obtaining the highest concentrations of glucose sugars with pre-treatment using sulphuric acid at 0. 5% and a time of fifteen minutes, and enzymatic activity in the hydrolysis,In order to show the influence of enzymatic activity on hydrolysis, a test was carried out using the test that gave the best result, in this case pre-treatment with 15 min and 0.5% $H_2 SO_4$, and the enzyme concentration was increased from 0.05FPU/g to 2.7 FPU/g and 7.0FPU/g, achieving 0.2 g of glucose/gram of bagasse with 7FPU/g, achieving a conversion of 83.5%. In the fermentation, the hydrolyzed liquor was used under the optimum conditions mentioned above, an industrial strain of *S. Cerevisiae was used,* and the medium was supplemented with $(NH_4)SO_4$ and $KH_2 PO_4$ as a source of ammonia and phosphorus. The best results came from experiments 8 (15g/L yeast, 0.9 g/L ammonium and 0.19 phosphorus) and 9 (10g/L yeast, 0.6g/L ammonium and 0.12 phosphorus), which produced 13 g/L of ethanol. Understanding the kinetics of the reaction will enable the design of enzymatic reactors, which could make the process cheaper and more viable. To this end, the material was initially characterized in order to assess the cellulose and lignin content. After this stage, the activity of the enzyme used was evaluated to confirm the manufacturer's data. Subsequently, the influence of substrate concentration on enzymatic hydrolysis was studied in order to apply the Michaelis-

Menten equation and calculate its constants. Subsequently, a CSTR was designed using the Michaelis-Menten equation as a reaction rate, in which a reactor of approximately 20 liters was proposed.

Keywords : Cashew, *Saccharomyces cerevisiae*, Pretreatment, Enzymatic Hydrolysis, Fermentation, Ethanol, Michaelis-Mente, CSTR.

SUMMARY

1. INTRODUCTION

Characterized by varied climatic conditions and different types, Brazil's extensive territory boasts extremely diversified agricultural production, which gives the country the title of the world's leading producer of various products. Although fruit-growing accounts for only around 5% of the country's cultivated areas, it is one of the activities capable of ensuring that Brazil has a significant percentage of the global production volume, placing it in first place in the ranking of fruit producers (CARRARO & CUNHA, 1994). Brazil's agricultural economy remains one of the most important in the world, with exceptional performance in the 2012/2013 harvest, when it reached a new record of more than 186.1 million tons and GDP growth of 9.7% in the first quarter of 2013 (ECON, 2013).

At the same time, millions of tons of agro-industrial waste are produced every year, most of which are disposed of in the environment, resulting in an excessive accumulation of organic matter in nature. Latin America produces around 500 million tons of agro-industrial by-products a year, with Brazil producing almost half of this amount (VARIZ, 2011). Among these agricultural products is cashew, which comes from the cashew tree (Anacardium occidentale L.), a tropical plant native to Brazil, scattered throughout most of its territory, especially in the northeast region, which accounts for more than 95% of national production. Its agro-industry produces approximately 276,399 tons of cashew nuts and 2 million tons of stalks per year (OLIVEIRA, 2008; IBGE, 2013).

The cashew chain comprises a set of activities that generate a large number of products, but the main commodity is the cashew nut kernel. However, it is estimated that more than 90% of the stalk is wasted, i.e. it is a by-product that is little used in the nut production chain (GUANZIROLI, 2009).

In view of the great waste of cashew stalks, there is growing scientific interest in studying the use of waste in general, seeking to present options for its use as a raw material to generate value-added products. Thus, cashew cultivation becomes a possibility for technological development, since cashew stalk waste, especially from the juice industry, can be reused to enrich animal feed (LUCIANO et al., 2011), produce pectinase enzymes (ALCANTRA; ALMEIDA; SILVA, 2010), produce ethanol and xylitol (LIMA et al., 2013; LIMA et al., 2014).

Lignin is a macromolecule made up of phenylpropane units, with a three-dimensional and amorphous conformation, representing 20 to 30% of the constitution of lignocellulosic materials. Hemicellulose is made up of short, linear and highly branched sugar chains. In contrast to cellulose (a polymer made up only of glucose), hemicellulose is a heterogeneous polymer made up of D-Xylose, D-Galactose, D-Glucose, D-Mannose and L-Arabinose.

The bioconversion of these polymeric materials requires a two-step process: acid or enzymatic hydrolysis of the sugar polymers into monosaccharides, followed by bioconversion of the monomers into a product of industrial interest. The hydrolysis methods are: acid, alkaline, enzymatic and without the use of catalysts. In the enzymatic hydrolysis process, lignocellulosic materials are hydrolyzed into fermentable sugars through the action of the cellulase enzyme, which is the enzyme capable of hydrolyzing the glycosidic bond between two or more carbohydrates or between a carbohydrate and a non-carbohydrate portion. The disadvantage of using enzymes to break down cellulose is that in addition to the need for pre-treatment to remove lignin, the cost of using them is high.

This work seeks to transform the energy potential available in cashew stalk waste through enzymatic processes into hydrated ethyl alcohol.

The sustainability and competitiveness of agribusiness necessarily involves reducing production costs. In various production systems, agro-industrial waste, when used rationally, reduces production costs, systematically benefits the environment and generates employment and income in the sector. With this in mind, the research seeks to invest in ingredients and alternatives that propose the transformation of these residues into socially and economically viable products in the cashew production chain in the Northeast. Ethanol is increasingly attracting the attention of researchers, companies and governments. This is due to the pressures and prospects of exhaustion of non-renewable sources of fossil fuels, as well as environmental concerns related to the emission of substances that compromise the environment.

Many studies of the kinetics of cellulose hydrolysis have been carried out, but the understanding of the dynamic interactions at the interface and the influence of these interactions on the reaction kinetics still remain inconclusive and limited. Several kinetic models have been satisfactory when it comes to predicting initial rates of cellulose hydrolysis reactions. Among the models proposed, the most widespread and simplest is that of Michaelis-Menten. It is important that a mathematical representation of a kinetic reaction takes into account information about the enzyme's catalytic system in order to cover all aspects of the reaction, but it is also essential that it is not too complex. Factors such as product inhibition, mass transfer, adsorption of the enzyme by the material, deactivation of the enzyme and biomass characteristics can all be taken into account.

1.1 Rationale

The cashew tree (*Anarcadium Occidentale L.*) is a medium-sized, xerophytic plant typical of tropical climates. It is a tree of Brazilian origin, more specifically from the northeastern coast, where it was spread by the colonizers to various countries in Africa and Asia.

On the national and international scene, the Brazilian market is a major player in the production of tropical fruits such as pineapple, orange, banana, acerola and others, and there is also a lot of space for cashew cultivation. However, there is a serious problem with this crop, due to the high commercial value added to the nut, where the pseudo-fruit (the stalk) is used for only 15% of Brazilian production. This is an intolerable waste for a region so lacking in food resources (EMBRAPA, 1999).

The use of cashew stalks to produce bioethanol is a way of using them and reducing the waste of this crop, providing a viable form of income for farmers throughout the Northeast, generating employment and helping to reduce environmental impact through renewable energy sources.

2. OBJECTIVES

2.1 General objective

To study the kinetics of enzymatic hydrolysis of the lignocellulosic raw material of cashew bagasse (*Anarcadium occidentale* L.) in order to produce hydrolyzed liquors for subsequent alcoholic fermentation (submerged fermentation) and to obtain bioethanol.

2.2 Specific Objectives:

- Physico-chemical characterization of the raw material cashew stalk bagasse (pH, apparent density, ash, moisture, total carbohydrates, cellulose, hemicellulose, lignin, TSS (°Brix), ART-total reducing sugars, reducing sugars-AR);
- Study of the kinetics of enzymatic hydrolysis of lignocellulosic matter (cashew bagasse), using adapted methodologies that are used for the hydrolysis of sugar cane bagasse and other lignocellulosic matter;
- Chemical and physicochemical characterization of enzymatic hydrolysates (liquors);
- Determination of kinetic parameters in alcoholic fermentations of hydrolysates (%conversion, productivity, $Y_{P/S}$).
- Determination of sugars in hydrolyzed liquors using RA analysis;
- Determination of the kinetic parameters of hydrolysis (V_0 , V_{max} , k_m) applied to the Michaelis-Menten equation;
- Designing an Ideal CSTR Reactor using the kinetic parameters of the Michaelis-Menten equation

3. LITERATURE REVIEW

3.1 Cashew

The cashew tree (*Anacardium occidentale L.*) is a medium-sized, hardy plant that belongs to the *Anacardiaceae* family and is typical of tropical climates. It is a genuinely Brazilian tree, from the northeastern coast, and was spread by the Portuguese colonizers, who were here, through the various countries of Africa and India (PARENTE *et al*, 1991). The true fruit of the cashew tree is the nut, which is made up of a shell (rich in phenols), in the innermost part of the nut is the kernel made up of two fleshy and oily cotyledons, which make up the edible part of the fruit. The fruit colloquially known as cashew is actually a pseudo-fruit (false fruit), corresponding to the hypertrophied floral peduncle with a color varying between yellow and red, the peduncle has developed in a different way, it is eaten fresh and appreciated for its juiciness, as shown in the figure below (FARIA, 1994). By weight, cashews are made up of 10% nuts and 90% stalks (GARRUTI, 2001).

Figure 1 Cashew - (a) Fruit (Chestnut), (b) Pseudofruit (Peduncle)

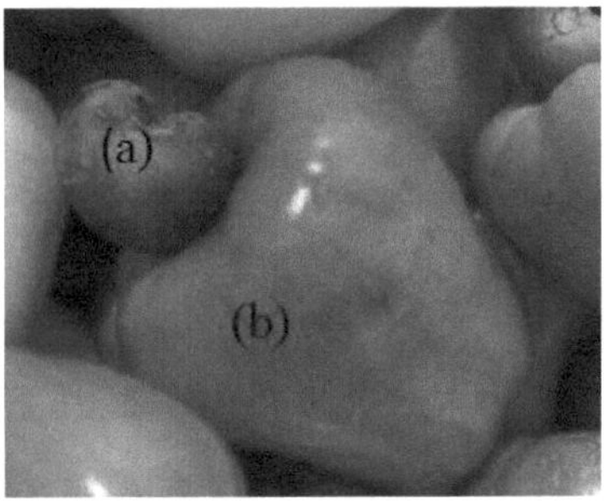

Source: Own authorship

Cashew is a crop of great economic importance for the Northeast region, both because it is consumed *fresh* and because its fruit is industrialized, resulting in juices and other products that are widely consumed on the domestic and foreign markets (PETINARI E TARSITANO, 2002). Large segments of the population in the Brazilian Northeast rely on cashew as an important source of income, and for many municipalities it is the main currency-generating crop (MENEZES E ALVES, 1995).

In Brazil, the area occupied by cashew trees corresponds to approximately 783,000 ha, with the Northeast region accounting for more than 90% of the harvested area and national production, with the states of Ceará, Piauí and Rio Grande do Norte accounting for around 85% of the cultivated area in the country according to data released by the IBGE for the 2009 harvest. It is estimated that only 15% to 20% of the pulp is used to make jams, juices, wine or for fresh consumption and 80% is wasted, meaning that thousands of tons of stalks are thrown away. A large part of the pseudo-fruit is also lost in the field when it is shelled for the nut processing industry (GARRUTI *et al.*, 2003).

The cashew stalk is a lignocellulosic material that has an estimated production in Brazil of around 1.8 million tons/year, basically concentrated in the Northeast region and with industrial use of only 15% of the total (GLOBO RURAL, 2005). In the Northeast, cashew nuts are practically used industrially to produce edible nuts and, on a small scale, juice and derivatives of the pseudo-fruit (EMBRAPA, 2004; Globo Rural, 2005). According to Holanda & Oliveira (2001), the amount wasted represents a high potential use for conversion by microorganisms.

The pomace, a by-product of the process of obtaining cashew juice, represents around 25% to 30% of the weight of the stalk and is generally discarded or used to make flour for animal feed (LEITE, 1994). Such a destination for the pomace can be considered a great waste, since this by-product of the cashew industry is a source of hydrolysable fibers. The hydrolysis of lignocellulosic biomass is essential for the production of fermentable sugars, which are then converted into ethanol by microbial action.

Bagasse, through biotechnological processes, can also be used to produce bioethanol and enriched protein (used in feed formulation). A large amount of bagasse is needed to produce ethanol (PIETROBON, 2008).

3.2 Biomass

Biomass is all non-fossilized organic matter of plant or animal origin that can be exploited for energy purposes. Biomass is thus a huge reservoir of energy, including forestry waste, agro-industrial waste and waste from the wood processing industry, urban waste, animal waste, etc., which can be converted into different forms of energy (heat, electricity, road fuels) (ADENE & AREAC). According to Moreschi (2005), biomass is the world's fourth largest source of energy, providing 14% of primary energy.

According to Rueda (2010) a potential source for low-cost ethanol production is the use of lignocellulosic materials, since there is a huge volume of agricultural waste that is not being fully utilized.

Lignocellulosic biomass is basically composed of cellulose, hemicellulose, lignin and small amounts of extractives and mineral salts (CARVALHO ET AL, 2005). Tsao (1978) states that in most lignocellulosic materials the existing ratio is 4:3:3, of cellulose, hemicellulose and lignin, respectively.

According to Pietrobon (2008), lignin, Figure 2, is a macromolecule made up of phenylpropane units, with a three-dimensional and amorphous conformation, representing 20 to 30% of the constitution of lignocellulosic materials. Hemicellulose is made up of short, linear and highly branched sugar chains. In contrast to cellulose (a polymer made up only of glucose, Figure 1), hemicellulose is a heterogeneous polymer made up of D-Xylose, D-Galactose, D-Glucose, D-Mannose and L-Arabinose.

Figure 2 Cellulose composition

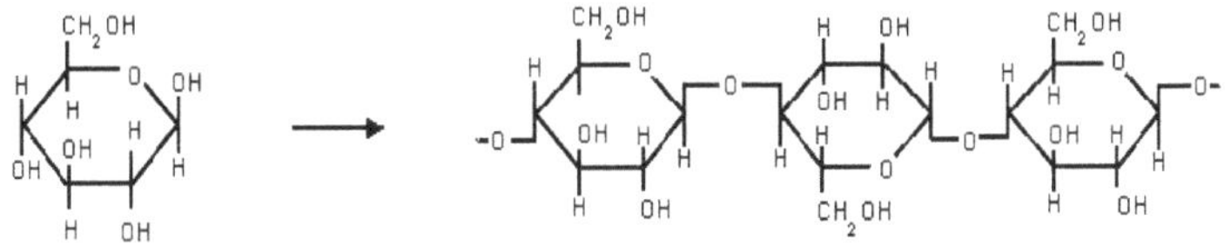

Source: Own Authorship

Figure 3 Lignin structure

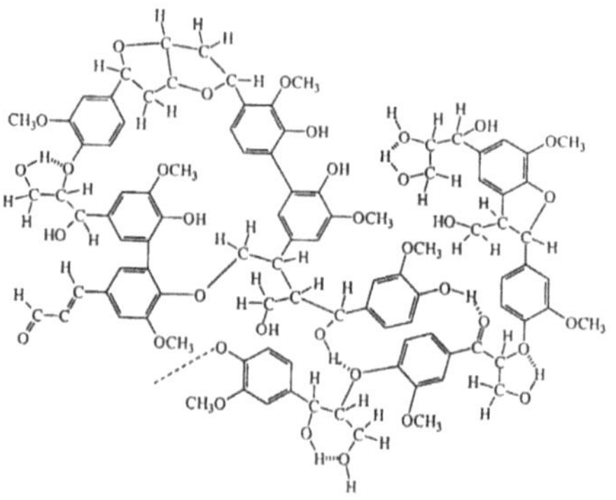

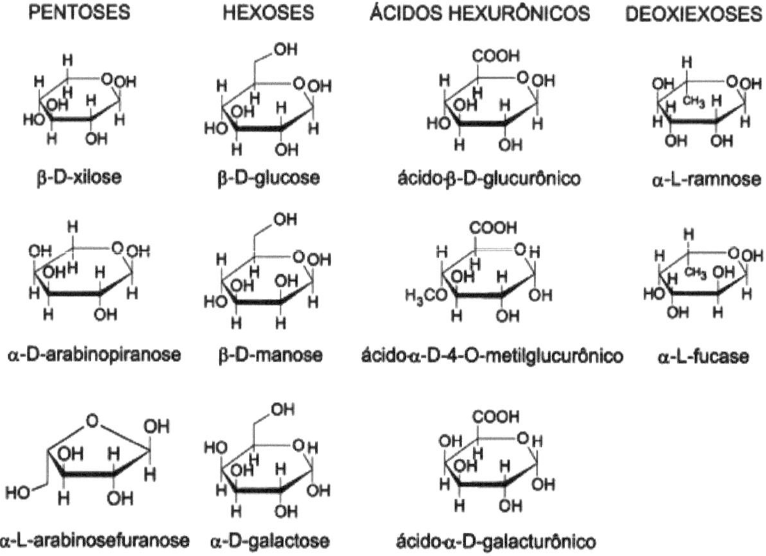

Source: (TAVARES, 2009).

Figure 4 Various sugar units

Source: (TAVARES, 2009).

3.3 Pre-treatment

In order to use agricultural waste, such as sugarcane bagasse, cashew nuts, among others, it is necessary to take steps to separate the lignocellulosic components. In order to obtain cellulosic ethanol, the separation of these constituents is carried out in four main stages: pre-treatment, enzymatic hydrolysis, fermentation and distillation (MAZIERO, 2010).

According to Guedes (2010), lignin is a structural compound and is integrated into the plant cell wall, providing impermeability, rigidity and protection against attacks on plant tissues. Therefore, lignin hinders the saccharification of cellulose. According to Silva (2010) there is a need for prior and adequate processing of these materials, making it necessary to break the cell wall (physical barrier), as well as reduce the crystallinity of cellulose and its association with lignin so that hydrolytic enzymes can access the biomass macrostructure in order to increase the yield of cellulose conversion into glucose. According to Salvachúa (2011) the aim of pre-treatment is to disrupt the structure of lignin and hemicellulose and improve enzyme accessibility.

Pre-treatment is any chemical, physical, physicochemical or biological treatment aimed at facilitating the enzymatic action of saccharifying biomass. This stage is the first in the study of the bioconversion of lignocellulosic compounds to ethanol. As a result of this phase, there is a significant increase in the yield of enzymatic hydrolysis and, with it, an increase in the concentration of fermentable sugars. The choice of pre-treatment conditions will depend on an overall assessment of the process, yield and maintenance of the physical structure of the inputs of interest (OLIVEIRA, 2010).

Various processes have been developed for the pretreatment of bagasse from lignocellulosic biomass, examples being steam explosion, hot water processes using dilute acid or even alkali and wet oxidation, and the use of Basidomycetes fungi. All these processes increase the digestibility of bagasse to some extent (ZHAO *et al* 2009).

3.4 Hydrolysis

According to Carvalho et al (2005), the use of forestry and agro-industrial waste as substrates in biotechnological processes for the production of high added value products is an attractive and promising alternative, since these materials are abundant, renewable and low cost. The bioconversion of these polymeric materials requires a two-step process: acid or enzymatic hydrolysis of the sugar polymers into monosaccharides, followed by bioconversion of the monomers into a product of industrial interest.

The hydrolysis methods are: acid, alkaline, enzymatic and without the use of catalysts. Acid saccharification is a mature technology, but it has the disadvantage of generating hazardous waste and technical difficulties in recovering the sugar from the acid. In addition to this disadvantage, Moreschi (2005) argues that acid or alkaline hydrolysis causes the decomposition of monomeric sugars, generating by-products such as hydroxymethylfurfural (HMF), which can interfere with future processes such as alcoholic fermentation. Enzymatic hydrolysis, on the other hand, is more efficient and proceeds under ambient conditions, without any generation of toxic waste (R UGGIERO, 2002). In the enzymatic hydrolysis process, lignocellulosic materials are hydrolyzed into fermentable sugars through the action of the cellulase enzyme, which is the enzyme capable of hydrolyzing the glycosidic bond between two or more carbohydrates or between a carbohydrate and a non-carbohydrate portion (BEGUIN, 1990).

Enzymatic saccharification has become a procedure with greater advantages over acid hydrolysis, because with the reduction in acid consumption it is possible to have both economic and environmental benefits (PIETROBON, 2008). While hydrolysis with concentrated acid is a faster and better-known method, the development of the process requires the use of corrosion-resistant materials, which tend to be more expensive.

Macromolecules of plant carbohydrates, such as cellulose, are mainly made up of their monomeric components by the formation of ether bridges, see Figure 1. Hydrolysis generally favors the breaking of cellulose's ether-ester bonds by adding one molecule of water for each broken bond. The acetyl groups in hemicellulose are connected to the pentoses by ester bonds: in the case of lignin, in addition to the ether bonds, C-C bonds also occur between the phenylpropane units (BOBLETER, 1994).

The cellulase complex involves at least three enzymes, which have very characteristic modes of action: β-1,4-glycano-glycohydrolase (an endoglycanase), β-1,4-glycano-cellobiohydrolase (an exoglycanase) and β-1,4-glucosidase (a cellobiase). Endoglycanase cleaves native cellulose at random, enabling the formation of smaller chains of cellotriose, cellobiose and glucose, while exoglycanase acts on the non-reducing ends of the chains, separating cellobiose molecules. Finally, cellobiase hydrolyzes cellobiose to glucose (REYES, 1998).

Hydrolysis can take place in the presence of acid catalysts (acid hydrolysis), alkaline catalysts (alkaline hydrolysis), enzymatic catalysts (enzymatic hydrolysis) or without the use of any catalyst (hydrothermolysis). According to Moreschi (2005) there are some disadvantages to the use of these processes, in the study of cellulose breakdown, using the acid route results in the decomposition of monomeric sugars produced during the reaction, occurring simultaneously with the hydrolysis of polysaccharides, but the use of bases to accelerate this reaction leads to the loss of hemicellulose which is not interesting in the procedure. The disadvantage of using enzymes to break down cellulose is that in addition to the need for pre-treatment to remove lignin, the cost of using them is high. The problem with using hydrothermolysis on lignocellulosic compounds is that around 20% of the glucose in the medium is degraded.

3.5 Enzymes

Enzymes are used as catalysts in many reactions. Their structure is largely made up of proteins and they can be linked to carbohydrates and lipids. Their most widespread form of use is the extracellular type of microbial origin.

The growth of enzyme production technology began in 1833 when French chemists Anselme Payen and Jean-Franois Persoz described the enzyme complex of germinating barley amylase, called diastase. After a lot of research, new enzymes were discovered, such as: pepsin (responsible for digesting albumin) in 1836, enzymes for cheese production in 1874, enzymes for converting starch into sugar (1894), malt and amylase in 1900, fungal amylase (1900), Pancreatic proteinases (1907), proteolytic enzymes used in brewing beer (1911), bacterial enzymes used in the textile industry, trypsin and pepsin (1950), pectinase (1950), papain (1950), *bacillus* proteinase in 1950, glucose isomerase (1976) among others. (PIETROBON, 2008)

A very widespread type of enzyme when it comes to converting lignocellulosic materials into glucose is cellulase. This enzyme is actually an enzyme complex produced by fungi or bacteria such as *Trichoderma reesei, Humicola insolens, Penicillium pinophilum* (fungi) and *Cellulomonas fimi, Clostridium thermocellum* (bacteria). The cellulase enzyme complex acts most efficiently on native or crystalline cellulose.

Cellulase causes the hydrolysis of cellulose, generating cellobiose, which in turn is transformed into glucose. This enzyme complex is made up of three main enzymes: endoglucanase, exoglucanase or cellobiohydrolase and β-glucose or cellobiase (PIETROBON, 2008).

Gilkes *et al.* (1991) states that cellulases are described as a complex group of enzymes with synergistic action:

Endoglycanases (EC 3.2.1.4): these are enzymes that catalyze the internal hydrolysis of bonds. Their natural substrate is cellulose and xyloglycan, with variable specificity for carboxymethyl cellulose (CMC), Avicel (crystalline cellulose), β-glycan and xylan.

Cellulose beta-1,4-cellobiosidase (EC 3.2.1.91): also known as exoglycanase, its catalytic action releases cellobiose from the non-reducing ends of the chains.

β-glucosidase (EC 3.2.1.21): known as gentobiase, cellobiase and amygdalase, it catalyzes the hydrolysis of non-reducing terminal β-D-glucose residues, releasing β-D-glucose.

β-1,3(4)-endo-glycanase (EC 3.2.1.6): also known as β-1,4-endoglycanase, β-1,3 endoglycanase or laminarinase, this enzyme catalyzes the internal hydrolysis generating glucose D-glycans. Its substrates are laminarin, lichenin and D-glycans.

The enzymes act synergistically to break down the cellulose. The endoglucanases act on the biomass in such a way as to break down its structure, creating ends which facilitate the action of the exoglucanases. The synergistic action ends when the cellulose chain is finally degraded to cellobiose. The end products of saccharification exert an inhibitory effect on the enzymes in such a way that exoglucanase is inhibited by cellobiose (competitive inhibitor) and β-glucosidases and some cellulases are inhibited by glucose.

The main problem in using enzymes for the hydrolysis of lignocellulosic materials is their cost, which is usually quite high, so the search for new enzymes with the same efficiency as those already used is the key to overcoming the expenses allocated to this process or, another possible way to reduce costs, is to reuse the enzymes, but it is known that, when reused, they lose their initial efficiency.(Garcia, 2009)

Cellulases have a wide variety of industrial applications, being used as an additive in the preparation of digestive enzymes, as a component of detergents, in the bleaching and softening of textile fibers, in the treatment of wastewater, in the food industry to increase the yield of starch and vegetable oil extraction and as animal feed additives (Amorin, 2010).

3.6 Enzyme kinetics

Many studies of the kinetics of cellulose hydrolysis have been carried out, but the understanding of the dynamic interactions at the interface and the influence of these interactions on the reaction kinetics still remain inconclusive and limited. There are some similarities between the studies, but there is still a great deal of divergence in the concepts related to the synergistic action of the different components of the cellulase enzyme, the effect of the structural change of cellulose during a continuous reaction and the mechanisms of inhibition of the enzyme by intermediate products of the reactions, such as cellobiose, and final products, such as glucose (GARCIA, 2009).

Moreira Neto (2011) states that the hydrolysis of cellulose, due to its heterogeneous nature, involves more steps than classic enzymatic kinetics. The main steps are described by:

1. Adsorption of cellulases onto the substrate via the binding domain ;

2. Location of a bond susceptible to hydrolysis on the substrate surface (end of chain if cellobiohydrolase, bond break if endoglycanase);

3. Formation of the enzyme-substrate complex (By binding at the end of the chain in the catalytic tunnel of the cellobiohydrolase, to initiate hydrolysis);

4. Hydrolysis of the β-glycosidic bond and simultaneous sliding of the enzyme along the cellulose chain;

5. Desorption of the cellulases from the substrate or repetition of step 4 or steps 2 and 3 only if the catalytic domain detaches from the chain;

6. Hydrolysis of cellobiose to glucose by the enzyme β-glucosidase (if present in the enzyme complex). In addition, product inhibition and changes in substrate properties throughout hydrolysis affect the above steps.

Various kinetic models have proved satisfactory when it comes to predicting the initial rates of cellulose hydrolysis reactions. Among the models proposed, the most widespread and simplest is the Michaelis-Menten model, which was originally developed for a system of homogeneous reactions. However, when considering the enzymatic hydrolysis of cellulose, this model does not represent the system in a real way, since it considers that the hydrolysis of bagasse is heterogeneous and does not take into account the multiple reactions that can occur in the reaction medium. If the aim is to predict the production of reducing sugars over long periods of time, there are discrepancies, mainly due to the evolution of the reaction caused by changes in the structure of the substrate and the loss of enzymatic activity (GARCIA, 2009).

The Michaelis-Menten equation measures the speed for an enzymatically catalyzed equation with a single substrate, at a fixed temperature and pH. It is the expression of the quantitative relationship between the initial velocity (V_0), the maximum initial velocity (V_m) and the initial substrate concentration ([S]), all related through the Michaelis-Menten constant (K_m). This constant is dynamic, or pseudo-equilibrium, and expresses the relationship between real steady-state concentrations rather than equilibrium concentrations. This equation can be described as follows: (MARIOTTO, 2006)

$$V_0 = \frac{V_m[S]}{K_m + [S]} \qquad (1)$$

The order of the constant K_m depends on two factors:
- When $[S] \ll K_m$, the reaction is first order.
- When $[S] \gg K_m$, the reaction is of zero order.
- When $[S]$ has intermediate values, the order of the reaction is between zero and one.

3.7 Ethanol production

3.7.1 Importance

Fifty years ago, the world consumed 4 billion barrels of oil a year and the average rate of discovery of new wells was 30 billion barrels a year. In recent years, 30 billion barrels a year have been consumed and discoveries have fallen to 4 billion barrels a year. In addition, another serious problem is associated with the use of oil, since its use as a fuel is responsible for 73% of CO production$_2$ (MAZIEIRO, 2010).

According to Vasconcelos (2010) there is a notorious concern about replacing fossil fuels, either because of their scarcity or because of environmental problems. Brazil has a wealth of knowledge in the area of biofuels, particularly in the use of sugarcane ethanol as an automotive fuel, and has one of the cleanest energy matrices in the world (OLIVEIRA, 2010). According to Rodrigues (2010) ethanol is used as a fuel on a large scale in Brazil, the United States and some European countries and is expected to be the renewable biofuel to dominate the transportation sector within 20 years. Almost all fuel ethanol is produced by fermenting sucrose in Brazil or corn glucose in the United States, but these raw materials will not be enough to satisfy international demand (ROSILLO-CALLE E CORTEZ, 1998).

The rapidly developing process of producing ethanol from lignocellulosic materials has immense potential for improving efficiency and cost. The commercialization of ethanol produced from lignocellulosic biomass is hampered mainly by the cost of currently available cellulase. Using cheap raw materials to produce cellulases could be a way of making ethanol production from lignocellulosic materials economically viable (Ruggiero, 2002). The production of ethanol from lignocellulosic materials by enzymatic hydrolysis is more efficient than acid hydrolysis (AMORIM, 2010).

The production of second-generation ethanol from lignocellulosic materials includes three main stages: (i) pre-treatment, (ii) enzymatic hydrolysis of cellulose or hemicellulose, and (iii) ethanol fermentation (SALVACHÚA ET AL 2011).

The benefits of biomass-to-ethanol technology have been previously indicated: increased energy security, i.e. new renewable energy sources; a reduction in greenhouse gas emissions, the foundation of a chemical process industry, macroeconomic benefits for rural communities and society as a whole (KNAUF AND MONIRUZZAMAN, 2004).

3.7.2 Fermentation

The ethyl alcohol agro-industry represents a considerable economic generator. As all alcohol production takes place via fermentation, knowledge of the fermentation process is essential and has been constantly improved (NOBRE *et al.*, 2007).

According to Corazza (2001) fermentation is a set of enzymatically controlled reactions through which an organic molecule is degraded into simpler compounds, releasing energy. According to Sousa (2010) alcoholic fermentation consists of the transformation of carbohydrates into ethanol and carbon dioxide through anaerobic metabolism, usually in the yeast *Saccharomyces cerevisiae*. However, other substances such as glycerol and acetic acid are produced in smaller quantities during this process.

Alcoholic fermentation (anaerobic catabolism) provides energy in the form of ATP or other energy transfer compounds for the biosynthesis of cellular material and the production of ethanol. These catabolic reactions occur with a large decrease in free energy, which together with the subsequent hydrolysis of ATP during the biosynthesis, transport and maintenance reactions, results in the production of heat (VOLP, 1997). A detailed diagram of the biotransformation of glucose into ethanol is shown in Figure 3.

According to Gutierrez (1990), the yeasts most commonly used to produce ethanol by fermentation in Brazil are *Saccharomyces cerevisiae* (baker's yeast). During alcoholic fermentation, yeasts produce not only ethanol and carbon dioxide, but also secondary compounds such as glycerol, higher alcohols, pyruvic and succinic acids, with glycerol being considered the most important component from a quantitative point of view. For ethanol production in distilleries, the formation of glycerol is undesirable as it reduces fermentation efficiency (GUITIERREZ, 1991).

The transformation of sucrose into ethanol and CO_2 by these yeasts involves 12 reactions in an orderly sequence, each catalyzed by a specific enzyme. This enzymatic apparatus is confined to the cell cytoplasm and it is therefore in this region of the cell that fermentation takes place (LEITE, 2010).

Figure 5 Diagram of the reactions in the biotransformation of glucose into ethanol

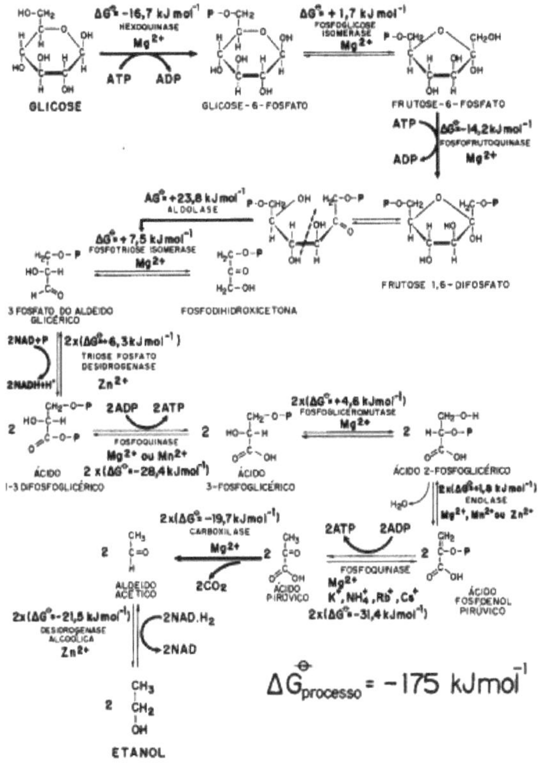

Source: (VOLP, 1997)

3.7.3 Distillation

According to Silva (2010), the main purpose of alcoholic distillation is to concentrate the fermented product to a high alcohol content, from a hydro-alcoholic mixture containing 6% to 10% ethanol by volume. Fermentate and distillate also contain various other volatile compounds in small quantities, usually derived from the fermentation process itself, which influence the quality of the final product. These compounds are methanol, higher alcohols (such as propanol, isopropanol, butanol, isobutanol, amyl, isoamyl), aldehydes (acetaldehyde, butyraldehyde, crotonaldehyde), organic acids (acetic, citric), ketones (acetone), esters (ethyl acetate, ethyl butyrate), ethers (acetal), etc.

From this impure and heterogeneous material, ethanol is separated by distillation, with varying degrees of purity and concentration. Distillation is a process for making the ethyl alcohol present in fermented liquids more concentrated. Through distillation, the fermented liquid is heated until it boils, the alcohol comes to the boil and its vapor, once condensed, forms a liquid with a higher alcohol concentration. To obtain industrial alcohol, the distillate from the first column is subjected to a second distillation in rectifying columns, whose function is to purify and concentrate the alcohol, obtaining a maximum alcohol content of 97.2% by volume (ABURANJA, 2009).

4. STATE OF THE ART

As mentioned above, there are four types of hydrolysis of lignocellulosic biomass. Acid, alkaline, enzymatic and hydrothermolysis (without the use of catalysts).

Santos et al (2008) when working with enzymatic hydrolysis of sugarcane bagasse with steam explosion pretreatment, obtained a maximum efficiency of 79.14% using 72.25 FPU/g of cellulase and minimum efficiency (46.91%) when using 14.45 FPU/g of enzyme was achieved, which, according to the authors, may be related to the lignin content in the substrate, showing the importance of using an adequate pretreatment.

Soares et al. (2008) studied the enzymatic hydrolysis of sugar cane bagasse, also using steam explosion pretreatment. Some conditions were varied: type of washing, grinding of the material and a cocktail of enzymes. The best results were obtained when the material was not ground, but washed with a 1% sodium hydroxide solution over 72 hours of reaction using 0.73g of Cellucast 1.5L and 0.33g of Novozym 188, both commercial cellulases. This test produced approximately 2.3g of glucose. The experiment that produced the least sugar was the one using the material washed with water, ground using 1.45g of Cellucast 1.5L and 0.33g of Novozym 188, in 72 hours of reaction producing one gram of glucose. According to the author, only the hemicelluloses are eliminated when washed with water, while washing with a NaOH solution causes the loss of lignin as well as hemicellulose. Therefore, in order to make better use of glucose, the authors recommend washing the material beforehand with an alkaline solution (NaOH).

Vasconcelos et al. (2010) studied the influence of pre-treatment with phosphoric acid on the enzymatic hydrolysis of sugarcane bagasse. The best results, with yields of cellulose in glucose above 80%, were obtained when acid treatment and a high temperature (186°C) were used beforehand.

Oliveira (2010), when studying the enzymatic hydrolysis of sugarcane straw pretreated with dilute sulfuric acid, found that as the pretreatment temperature increased, hemicellulose was gradually removed. As far as lignin is concerned, the removal was milder as the temperature increased, reaching a maximum yield of 54%. In order to remove greater amounts of lignin, the author used alkaline delignification, obtaining yields of up to 76%. In its enzymatic hydrolysis, a maximum glucose concentration of 59 g/L was achieved.

When studying the enzymatic hydrolysis of rice husks, Reyes (1998) used four types of pre-treatment: ultraviolet irradiation, hydrogen peroxide, ozone and sodium chlorite. The best delignification results were obtained with pre-treatment with sodium chlorite, but greater glucose production was achieved when pre-treating with ozone. For the author, treatment with sodium chlorite, although causing the most significant changes with regard to lignin, proved to be detrimental to hydrolysis, perhaps due to obtaining substrates that partially inhibit cellulases.

Caldas (2010) researched the acid hydrolysis of sugarcane bagasse using hydrochloric acid. Lewis acid (lithium chloride) was used to increase the catalysis of the reaction. In the temperature range studied, 140-180ºC, temperature was not a major parameter in the reaction, once again indicating the decisive effect of the Lewis acid. In this process, the production of xylose was equivalent to the production of glucose, at 25 g/L. The conversion of cellulose into glucose was not significant, as the decision was made to work under mild conditions.

According to Silva (2010), when he worked on the acid hydrolysis of the giant variety of forage palm, the best glucose results were achieved using 50g of biomass, a temperature of 121ºC, 100mL of 1% H_2SO_4 and a reaction time of 30 minutes, achieving 3.72 g/L of glucose after hydrolysis. Their results show that palm can be a biomass with the potential to provide fermentable sugars.

Carvalho et al (2005) worked with acid hydrolysis using sulfuric acid and eucalyptus chips as biomass, obtaining xylose ($24.32g.dm^{-3}$), glucose ($1.53g.dm^{-3}$) and arabinose ($0.53g.dm^{-3}$), the former being the predominant sugar in the hydrolysate.

Moutta (2010) analyzed that sugarcane straw, after acid hydrolysis with diluted H_2SO_4, has 8.56 g/L of glucose, 59.15 g/L of xylose, 8.78 g/L of arabinose, 1.43 g/L of acetic acid, 0.03 g/L of furfural and 0.002 g/L of hydroxymethylfurfural. After this stage, alcoholic fermentation was carried out with the microorganism *Pichia stipitis*, obtaining an ethanol concentration of 24.74 g/L after sixty hours of fermentation.

Studies carried out by Martín et al (2007) compared the efficiency of pre-treatment under different conditions (temperatures between 205-215ºC, times between 5 and 10 minutes, impregnation with sulphur dioxide and sulphuric acid) and subsequent hydrolysis of sugar cane bagasse. The parameters identified were the production of fermentation-inhibiting compounds and the production of total sugars. Among the experiments, it was found that the fermentation of bagasse pretreated at 205ºC and 10 minutes resulted in the highest ethanol production per gram of bagasse, double that obtained at 215°C and 10 minutes. It was also observed that bagasse impregnated with sulphur dioxide before thermal pre-treatment led to high sugar production.

Koo et al (2011) studied the production of ethanol from wood (Liriodendron tulipifera) using alkaline organosolv pretreatment with 50% ethanol and 1% sodium hydroxide (w/w) at different times. After analyzing the pretreatment followed by enzymatic hydrolysis, it was observed that the highest yield was obtained at 140°C. The saccharification and fermentation process was carried out simultaneously with the pre-treated material at this temperature, reaching the highest glucose concentration at 24 hours, where the start of fermentation was noted, with the highest ethanol concentration at 96 hours, which was 30 g/L .

According to Salvachúa et al (2011), the bioprocessing of wheat straw with basidomycetes to produce ethanol is little covered in the literature, and this process is far from being used for industrial purposes for various reasons such as time, but this procedure can be combined with other more effective methods. Twenty-one species of basidomycetes were used in this study to investigate which would be the most effective. The microorganisms that performed best were *I. Lacteus* and *P. subvermispora*, the former managed to increase glucose by 66% and the latter by 69% after 21 days of incubation. These pre-treatments were then fermented with *S. cereviciae* and both achieved 62% conversion of glucose into ethanol.

Harun et al (2011) studied the effect of alkaline pretreatment of microalgae of the species *Chlorococcum infusionum* at different temperatures, times and concentrations of alkali. As microalgae have no lignin, it is assumed that during alkaline pre-treatment, the fermentable sugars stored within the cell wall are released and are available to be converted into bioethanol. The highest sugar release (350.13 mg/g) was obtained when the biomass was treated with 0.75% w/v NaOH at 120 °C for 30 minutes. In this test, when fermented with *S. cereviciae, the* highest ethanol production was obtained, with a yield of 26.16 g ethanol/g microalgae.

According to Yoshikuni et al (2008), the use of acid pretreatments is the most popular form of hydrolysis, but the use of strong acids can cause damage to industrial equipment. The corrosion rate of 0.25% sulfuric acid, for example, is 1.25mm/year at boiling temperature, while that of 0.20% acetic acid is 0.125mm/year. Based on this, the author looked for a pre-treatment method that didn't use H_2SO_4 . The author used a water/ethanol/acetic acid treatment for eucalyptus chips, using enzymatic hydrolysis, where the highest conversion of cellulose into glucose was 113.4%, using 200°C and a 3:1 ethanol/water ratio as pre-treatment conditions, with 1% acetic acid at 60 min. According to the authors, this conversion was more than one hundred percent, because in addition to cellulases, there were hemicellulases in the enzyme complex, meaning that cellulose was converted into glucose as well as hemicellulose.

Santos & Gouveia 2009 studied the influence of lignin on ethanol production by enzymatic hydrolysis. The study was carried out with sugarcane bagasse pretreated by steam explosion (A), and sugarcane bagasse pretreated in the same way, only delignified downstream (B). The production of glucose was 44.11 g/L in (A) and 45.65 g/L (B). After this stage, the hydrolyzed materials were fermented and (B) (10.5 g/L of ethanol) produced around 10% more ethanol than (A) (9.5 g/L of ethanol). It can therefore be concluded that the presence of lignin in the hydrolysis of non-delignified material was a limiting factor for enzyme access, and that delignification increased ethanol yields by 10% in relation to glucose and 96% in relation to bagasse.

5. MATERIALS AND METHODS

5.1 Workplace

This work was carried out at the Biochemical Engineering Laboratory in conjunction with the Laboratory of Particulate Systems and Porous Media, both belonging to the Chemical Engineering Academic Unit at the Science and Technology Center of the Federal University of Campina Grande, PB.

5.2 Raw materials

The raw material used was the cashew stalk residue from *Anarcadium Occidentale L.* The cashew was purchased at different times of the year from open-air markets in the city of Campina Grande /PB. In the laboratory, the stalk was pressed to obtain juice and bagasse, which was dried in a circulating oven for 48 hours and then the material was ground and stored until it was ready for use.

5.3 Characterization of the raw material

5.3.1 Apparent density

100 grams of the material were weighed and placed in a cylinder to determine the volume occupied without compaction.

$$Densidade\ aparente = \frac{massa, g}{volume\ ocupado, cm^3} \quad (2)$$

5.3.2 pH

A suspension was prepared with 10mL of distilled water and 0.5g of the solid sample. After homogenization, the suspension was left to stand for 30 minutes, then the pH was measured in a digital potentiometer, previously calibrated with the standard solutions (Brasil, 2005).

5.3.3 Humidity

For the moisture analysis, 3 to 5 grams of the sample were weighed into a porcelain crucible that had been previously dried and weighed. The crucibles were placed in an oven at 105°C for 24 hours.

$$Umidade,\% = \frac{(peso\ inicial - peso\ final\ da\ amostra)}{peso\ inicial\ da\ amostra} \times 100 \tag{3}$$

5.3.4 Ash

Empty porcelain crucibles were placed in the muffle furnace and left at 550°C for 15 minutes. They were then left in a desiccator until they reached room temperature and weighed empty and with one to two grams of the sample. After weighing, they were taken to the muffle furnace for five hours at 550ºC until they were light gray. They were then left in the desiccator until room temperature and weighed again (Brasil, 2005).

$$Cinzas,\% = \frac{peso\ final\ da\ amostra}{peso\ inicial\ da\ amostra} \times 100 \tag{4}$$

5.3.5 Soluble solids content (ºBrix)

The concentration of soluble solids measured in ºBrix is a measure of the amount of sugar present in the sample. 9 ml of distilled water was added to 1g of the residue, stirred until perfectly homogenized and the suspension left for 30 minutes. After this period, the suspension was filtered with absorbent cotton and read on a refractometer, the result being multiplied by ten, due to the dilution, to determine the soluble solids content of the residue (Brasil, 2005).

5.3.6 Reducing sugar (RA) content

The DNS (3,5-dinitro salicylic acid) method, Miller (1959), is based on the reduction of 3,5-dinitro salicylic acid to 3-amino-5-nitrosalicylic acid, concomitantly with the oxidation of the sugar's aldehyde group to a carboxylic group. After heating, the solution turns reddish and is read on a UV spectrophotometer at 540 nm, according to the Embrapa CNPAT procedure.

5.3.7 Total reducing sugar content (ART)

The method for obtaining ART is based on the reduction of DNS (3,5-dinitro salicylic acid) to 3-amino-5-nitro salicylic acid, concomitantly with the oxidation of the sugar's aldehyde group to a carboxylic group.

$$ART\left(\frac{gART}{g\ amostra}\right) = \frac{ABS*curva*f.d.*V}{g\ amostra*10} \tag{5}$$

ABS = absorbance

Curve = calibration curve, obtained by doing the same procedure but using glucose solution

d.f. = dilution factor (5)- (1ml sample + 1ml HCL + 3ml NaOH)

V = volume used to prepare the sample dilution (in this case 100ml)

10 = conversion factor.

5.3.8 Determination of cellulose and lignin content

The cellulose content of the materials used was determined according to Updegraff's methodology (1969) for ground sugarcane bagasse. The material was incubated at 100 °C with nitric acetic reagent (80% acetic acid and concentrated nitric acid) for 30 minutes. After centrifugation at 3000 x g for 60 minutes at room temperature, 72% sulfuric acid was added to the precipitate. To determine the cellulose content, 8.0 mL of 2% anthrone solution and 4.0 mL of distilled water will be added to 1.0 mL of the sample, and the solution will be incubated in boiling water for 15 minutes. The solution will be analyzed in a spectrophotometer at 620nm. A cellulose standard curve (mg/mL) will be made and analyzed in a UV spectrophotometer at 620nm. The lignin content will be determined according to the methodology of Milagres et al. (1994), considering as lignin the solid residue remaining after hydrolysis with 72% sulfuric acid and post-hydrolysis with 4% sulfuric acid. The residue will be filtered and washed thoroughly with distilled water until the excess acid is removed, and kept in an oven at 105 °C until it reaches a constant mass.

Hemicellulose

The hemicellulose fraction is determined by the difference between the fraction of N.D.F. (Neutral Detergent Fiber) and A.D.F. (Acid Detergent Fiber). The hemicellulose fraction is a group of substances which includes pentose polymers (xylose, ribose, etc.) and certain polymers of hexoses and uranic acids, according to Silva (1998).

5.4 Determination of Enzymatic Activity

The cellulolytic activity analyzed in this work was endoglycanase, also called carboxymethyl cellulase (CMCase), following the procedure described in Menezes (2009), which is based on the ability of the enzyme extract to release reducing sugars in the presence of carboxymethyl cellulose due to the hydrolysis of the cellulose source into glucose.

For the analysis, 1 mL of the enzyme extract was placed in a test tube with 1 mL of a 1% (w/v) carboxymethylcellulose solution, prepared in 50 mM sodium acetate buffer and pH 5.0, at a temperature of 50°C for exactly 60 minutes. After this time, 1 mL was removed from the test tube and placed in another test tube with 1 mL DNS to determine the concentration of reducing sugars, which followed the methodology of Miller (1959) as described in the physicochemical characterization of washed and unwashed cashew bagasse. A blank sample was taken for each analysis, and instead of placing 1 mL of the enzyme extract with 1 mL of the carboxymethylcellulose solution, 1 mL of the enzyme extract was placed with 1 mL of 50 mM sodium acetate buffer, pH 5.0. This determined the concentration of reducing sugars in the enzyme extract without the presence of carboxymethylcellulose (ARSwithout CMC) and the concentration of reducing sugars in the enzyme extract after incubation with the cellulose source (ARWith CMC).

A unit of enzyme activity (U) was defined as the amount of enzyme capable of releasing 1 μmol of glucose per minute at 50°C. In this work, enzyme activity was expressed as U/g, which was calculated following this equation.

$$CMCase(U/g) = \frac{AR_{Liberado}x10^6}{180x60} = \frac{\left(AR_{com\,CMC} - AR_{sem\,CMC}\right)x10^6}{180x60} \quad (6)$$

5.5 Pre-treatment

The pre-treatment was carried out with sulphuric acid and phosphoric acid (separately), in order to remove hemicellulose and lignin from the bagasse. The reactions were conducted in a stainless steel reactor from MAINTEC FORNOS INTI with a FE50RP thermal controller, the concentrations of the acids were variable and so was the reaction time, the value of which can be found in the experimental design matrix in the following item, the pre-treatment temperature was 120ºC.

5.6 Hydrolysis

The pre-treated bagasse was hydrolyzed in an incubator (Marconi MA-420), shaken at 150 rpm. The enzymatic hydrolysis was carried out in 250 mL erlenmeyer flasks at a temperature of 50°C and pH 4.8 (citrate buffer). One gram of pre-treated bagasse was hydrolyzed with a commercially available enzyme, cellulase produced by *Aspergillus niger* C1184-5KU (Sigma-Aldrich). They were placed in a 250 mL erlemnayer, together with 15 mL of buffer solution (2g citric acid/7g sodium citrate; pH between 4.8 and 5.0), adding 2 g of pre-treated cashew bagasse. The bagasse was then placed in a shaker under the following conditions: rotation of 150 rpm, temperature of 50 °C, and a time of 48 and 72 hours. The acid concentrations in the pretreatment, the amount of enzyme (cellulase) and the reaction time were defined as the variables at the two levels (-1 and +1) studied in the hydrolysis process for each input variable. The acid concentrations were set at 0.5, 2.0, 3.5%, the amount of cellulase at 3, 6, 9mg and the reaction time at 15, 97, 180 minutes. Table 1 shows the actual and coded values for the factorial design.

Table 1Actual values and factor levels of the full factorial design (2^3) studied.

Factors	-1	0	1
%Acid	0,5	2,0	3,5
Cellulase (mg)	3	6	9
Time(min)	15	97	180

This factorial design requires at least twenty-two experiments, i.e. eleven for the pre-treatment with phosphoric acid and eleven for the pre-treatment with sulphuric acid, which are shown in Table 2.

Table 2Design of experiments matrix in coded and real form

Experiment	% Acid	Time (min)	Quantity of Cellulase (mg)

1	- (0,5)	- (15)	-3
2	+ (3,5)	- (15)	-3
3	- (0,5)	+ (15)	-3
4	+ (3,5)	+ (15)	-3
5	- (0,5)	- (180)	+6
6	+ (3,5)	- (180)	+6
7	- (0,5)	+ (180)	+6
8	+ (3,5)	+ (180)	+6
9	2	0 (97)	9
10	2	0 (97)	9
11	2	0 (97)	9

5.7 Enzymatic Hydrolysis Kinetics

The pre-treated bagasse was hydrolyzed in an incubator (Marconi MA-420), stirred at 150 rpm. The enzymatic hydrolysis was carried out in 250 mL erlenmeyer flasks at a temperature of 50°C and pH 4.8-5.5 (citrate buffer). One gram of pre-treated bagasse was hydrolyzed with a commercially available enzyme, cellulase produced by *Aspergillus niger* C1184-5KU (Sigma-Aldrich). They were placed in a 250 mL erlemnayer, together with 15 mL of buffer solution (2g citric acid/7g sodium citrate; pH between 4.8 and 5.0). The bagasse was then placed in a shaker under the following conditions: rotation of 150 rpm, temperature of 50 °C, enzyme concentration of 50 U/g (Units of endoglucanases per gram of bagasse) and samples were retained at 1, 3, 5, 10, 30, 45, 60, 120 minutes.

5.8 Characterization of hydrolysis sugars

The sugar content was determined using high-performance liquid chromatography (HPLC) equipped with a ProStar 210 pump (Varian); a manual injector with a 20μL loop; and a ProStar 356 refractive index detector (Varian). Hi-Plex H stainless steel analytical column (300mm x 7.7mm; Varian), and the conditions of the

Operations were as follows: Column temperature 40°C; Mobile phase: milliQ water at a flow rate of 0.6 mL/min; Analysis time: 15 for sugar content. Internal standard solutions of sugars: glucose, xylose, arabinose and sucrose (Sigma 99.99% HPLC grade) were used to quantify the components of the liquor.

5.9 Design of an Ideal CSTR for Cashew Bagasse Hydrolysis

Applying a mass balance to a CSTR reactor, assuming an ideal mixing reactor, we write:

$$F_{A0} - F_A + \int r_a dV = \frac{dN_A}{dt} \qquad (7)$$

Assuming steady state, ideal reactor and reaction rate independent of volume:

$$F_{A0} - F_A - r_a V = 0$$

$$V = \frac{F_{A0} - F_A}{-r_a} \qquad (9)$$

$$F_A = (1 - X_a)F_{A0} \qquad (10)$$

The reaction rate is known as the initial velocity of the Michaelis-Menten equation, described by equation (1):

$$-r_a = V_0 = \frac{V_m C_A}{k_m + C_A} \qquad (11)$$

$$V = \frac{F_{A0} - (1 - X_a)F_{A0}}{\frac{V_m C_A}{k_m + C_A}} \qquad (12)$$

$$V = \frac{F_{A0} X_a (k_m + C_A)}{V_m C_A} \qquad (13)$$

Where equation (12) describes the design equation of a CSTR for enzymatic hydrolysis, where the reaction rate is the Michaelis-Menten equation.

5.10 Ethanol production

5.10.1 Fermentation

Fermentation kinetics studies for ethanol production will be carried out in 250ml erlenmeyer flasks operating in batches in an incubator (Marconi MA-420), stirred at 150 rpm at 30ºC, using the *Saccharomyces cerevisiae* yeast selected and activated for the cepq distillery model LNF FT-858. The yeast will be incubated in the multiplication process, before inoculation, to obtain 10^8 cells/mL, using a biological reactor aerated with a sucrose solution and supplemented with phosphorus and ammonia. From the hydrolysates (liquors) characterized, the following steps will be carried out: Wort correction Nitrogen and phosphorus-based nutrients will be added to regulate the wort, providing favorable conditions for the development of microorganisms. The pH will be corrected to around 4.0. The kinetic study of alcoholic fermentation will analyze the concentration of glucose and ethanol at 0 2 6 8 12 24 36 48 72 hours, which will be quantified by HPLC.

According to Silva (2008), the must should be supplemented with ammonium sulphate as a source of nitrogen and $KH_2 PO_4$ as a source of phosphorus. The concentration of nitrogen, phosphorus and yeast was varied using a 2^3 experimental design according to Table 3.

Table 3Experimental Design for Fermentation

Factors	-1	0	+1
Yeast concentration (g/L)	5	10	15
Concentration $(NH)_{42} SO_4$ (g/L)	0,3	0,6	0,9
Concentration $KH_2 PO_4$ (g/L)	0,05	0,12	0,19

With this experimental planning, at least eleven experiments are required, which are described in Table 4.

Table 4Experiment design matrix in coded and real form

Experiment	Yeast (g/L)	$(NH_4)SO_4$ (g/L)	$KH_2 PO_4$ (g/L)
1	5(-)	0,3(-)	0,05(-)
2	15(+)	0,3(-)	0,05(-)
3	5(-)	0,9(+)	0,05(-)

4	15(+)	0,9(+)	0,05(-)
5	5(-)	0,3(-)	0,19(+)
6	15(+)	0,3(-)	0,19(+)
7	5(-)	0,9(+)	0,19(+)
8	15(+)	0,9(+)	0,19(+)
9	10(0)	0,6(0)	0,12(0)
10	10(0)	0,6(0)	0,12(0)
11	10(0)	0,6(0)	0,12(0)

5.10.2 Inoculation

Inoculation will take place with the addition of the microorganism. Firstly, yeast of the genus *Saccharomyces cerevisiae* will be used (commercial baker's yeast with 70% humidity from the Fleschamm brand). The fermentation will be monitored at regular intervals in order to define the fermentation profile of the production, thus obtaining the kinetic parameters (kinetic study) and adjusting the data using models from the literature (Monod, Levenspiel).

6. RESULTS AND DISCUSSION

6.1 Characterization

The physicochemical composition was carried out on the dried cashew bagasse *in natura* (Table 5).

Table 5Characterizations of cashew bagasse

Analysis	Unit	
pH		$4,76 \pm 0,05$
Humidity	%	$7,26 \pm 0,20$
Cellulose	%	$34,93 \pm 0,55$
Lignin	%	$23,53 \pm 0,37$
Extractives	%	$15,13 \pm 0,005$
Soluble solids	%	0

The main component of free soluble solids in this material is sugar, which was not found in this bagasse, as the bagasse was washed to remove all free glucose.

Looking at the data described in Table 5, it can be said that: there is a large amount of lignin, so it is clear that pre-treatment must be used before hydrolysis. There is also a high concentration of cellulose, so this material could be raw material for the production of fermentable sugars via the enzymatic hydrolysis process. Cassales (2010) quantified the lignin present in soybean hulls and found a value of 9.1%, well below that found for cashew bagasse.

6.2 Enzymatic Activity

Initially, the activity of the enzyme used in the enzymatic hydrolysis process of the lignocellulosic material, C1184-5KU (Sigma-Aldrich) produced by the microorganism *Aspergillus niger*, was analyzed in order to find out how much should be used in saccharification, using the methodology in item 4.5. According to this procedure, the enzyme activity is 1172.54 U/g. According to the information provided by the manufacturer, the activity would be 1240 U/g. This result does not differ much from the experimental result, i.e. the enzyme was not degraded or deactivated by conservation.

6.3 Enzymatic hydrolysis

According to the experimental design matrix shown in section 5.7, it was possible to carry out the pre-treatment experiments followed by enzymatic hydrolysis. The pre-hydrolysis (pre-treatment) data can be found in Tables 6 and 7, for pre-hydrolysis with sulphuric and phosphoric acid respectively.

Table 6Glucose concentration after 48 hours of hydrolysis with sulfuric acid pretreatment.

Time (min)	Acid concentration (%)	Enzyme concentration (FPU/g bagasse)	Glucose concentration in 48 hours (mg/L)	Glucose concentration in 48 hours (g glucose/g bagasse)
	0,5	0,05	2475,57	0,0201
15		0,1	742,35	0,0060
	3,5	0,05	134,52	0,001
		0,1	1846,35	0,0150
		0,75	223,63	0,0018
97	2,0	0,75	173,39	0,0014
		0,75	108,64	0,0009
	0,5	0,05	1288,2	0,0104
		0,1	1430,18	0,0116
180	3,5	0,05	182,52	0,0015
		0,1	259,47	0,0021

Table 7Glucose concentration after 48 hours of hydrolysis with phosphoric acid pretreatment

Time (min)	Acid concentration (%)	Enzyme concentration (FPU/g bagasse)	Glucose concentration in 48 hours (mg/L)	Glucose concentration in 48 hours (g glucose/g bagasse)
15	0,5	0,05	138,86	0,0011
		0,1	696,4	0,0056
	3,5	0,05	128,57	0,0010
		0,1	352,67	0,0029
97		0,75	208,22	0,0017
	2,0	0,75	140,34	0,0011
		0,75	190,48	0,0015
180	0,5	0,05	424,19	0,0034
		0,1	517,29	0,0042
	3,5	0,05	295,25	0,0024
		0,1	678,58	0,0021

It is clear that pre-treatment with sulphuric acid was more efficient than with phosphoric acid, as the highest concentration in acid pre-hydrolysis was 0.0201 g/g and that of pre-hydrolysis with phosphoric acid was around four times lower, 0.0056g/g. It can also be seen that the best result, Table 7, was achieved with the lowest acid concentration and reaction time. Oberoi (2010) explains that as the acid concentration in the pre-treatment increases, the glucose present in the biomass degrades. These results are in agreement with Pietrobon (2008) when he studied pre-treatment with sulphuric acid followed by enzymatic hydrolysis for sugarcane bagasse.

The current experiment was carried out at 48 and 72 hours, but only the first time is shown here, as there was a decrease in glucose concentrations. This may be explained by the degradation of the sugar during the reaction, or the enzyme may have been inhibited by the product formed.

Yoshikuni (2008), when working with eucalyptus chips from ethanol/water pretreatment, also obtained better results with 48 hours of hydrolysis. Harun (2011) obtained 350 mg/g of glucose in microalgae with 30 min of alkaline hydrolysis (NaOH) 0.75%(w/v) and 120 °C, but this time was short because microalgae do not have lignin in their cellular composition, facilitating hydrolysis.

In order to show the influence of the enzyme concentration on the hydrolysis process, a test was carried out using the test that gave the best result, in this case pretreatment with 15 min and 0.5% $H_2 SO_4$, and the enzyme concentration was increased from 0.05FPU/g to 2.7 FPU/g and 7.0FPU/g. The results are shown in Table 8.

Table 8Influence of enzyme concentration (cellulase) on hydrolysis

Time (min)	H concentration $_2 SO_4$ (%)	Enzyme concentration (FPU/g)	Glucose concentration at 48 hrs (mg/L)	Glucose concentration at 48 hrs (g/g)
15	0,5	2,7	13236,84	0,1075
15	0,5	7,0	24500,00	0,2000

Table 8 shows the influence of enzyme concentration on hydrolysis, as there was a significant tenfold increase when the enzyme concentration in the medium was 7.0 FPU/g. Analytical analyses of the constitution of *fresh* cashew bagasse were carried out and showed that this material contains approximately 34.0% glucose (0.34 g glucose/g dry bagasse), 15.1% hemicellulose and 23.7% lignin. Thus, with cellulase activity of 2.7 FPU and 7.0 FPU, the glucose release yields were 45% and 83.5% respectively.

The subsequent study of the alcoholic fermentation of this hydrolyzed liquor will be carried out under the conditions maximized in the current study, i.e. treatment of the bagasse with 0.5% sulphuric acid for 15 minutes and hydrolysis of the treated bagasse in 48 hours with the addition of cellulase of 7.0-10 FPU/g dry bagasse.

6.4 Fermentation

The kinetic data was also evaluated through the profiles of glucose and ethanol as a function of time, which can be seen in Figures 6 and 7. It was decided to present the profiles with the concentrations of sugars and ethyl alcohol given in g/L. The kinetic data values in g/L can be found in the table in Appendix A.

Figure 6 Decay of glucose concentration with time in fermentation

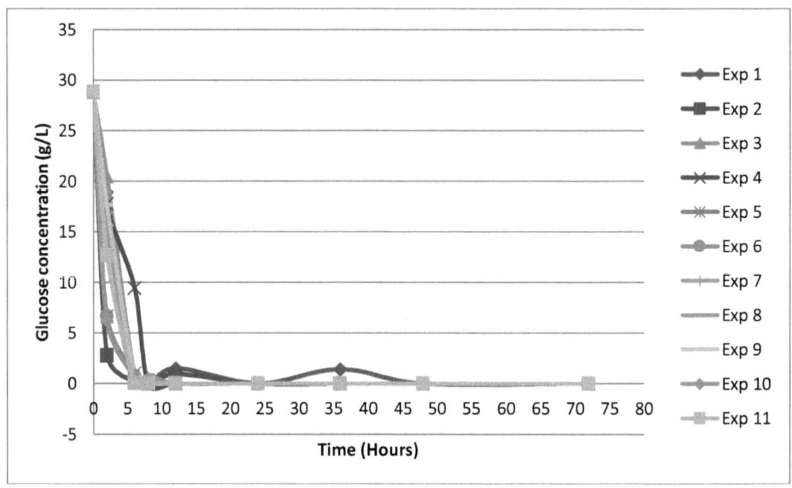

Source: Own authorship

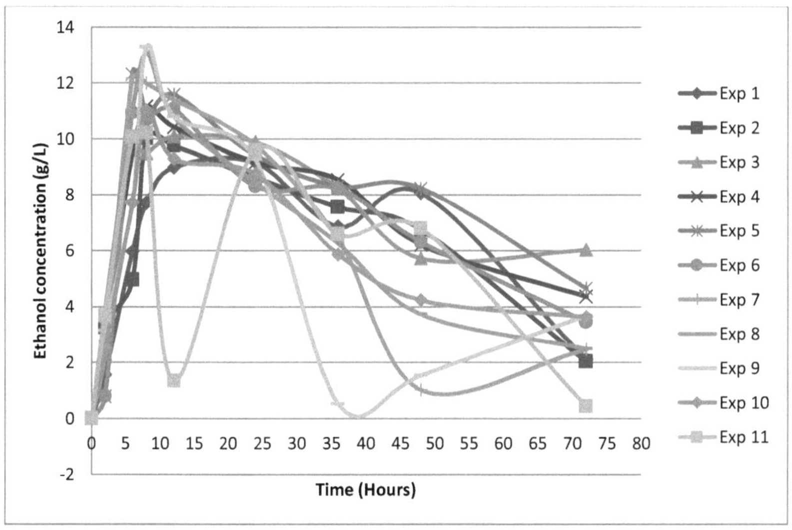

Source: Own authorship

Looking at Figure 6, it can be seen that after 8 hours of processing, all the sugar is consumed and remains so until the end of fermentation.

Figure 7 shows that, approximately, between 8-12 hours all ethanol is formed, reaching up to 13 g/L (experiments 8 and 9), after which it decreases in all cases, for example experiment 11 reaching almost zero concentration, for this may have occurred a contamination by bacteria found in the air such as those of the genus *Acetobacter* or *Gluconobacter* that convert ethanol into acetic acid, this is evidenced in the data presented in Table 9.

Oberoi (2010), when working with fermentation of acid hydrolysate from orange pomace, observed the same behavior of a decrease in ethanol concentration after reaching its peak.

Rocha (2010) obtained 20 g/L of ethanol using cashew bagasse pretreated with NaOH and hydrolyzed ezymatically, and without adding any supplements to the fermentation, obtaining a productivity of 3.33 g L h^{-1-1} .

Table 9Growth of Acetic Acid Concentration in Fermentation in Experiment 5

Time (hours)	Glucose (g/L)	Ac. Acetic Acid	Ethanol (g/L)
0	28,843	-	0
2	6,699	-	0,782
6	1,043	-	12,313
8	0	-	10,860
12	0	-	11,570
24	0	2,192 (78,3%)	9,416
36	0	3,900 (93,5%)	8,306
48	0	4,200 (96,7%)	8,230
72	0	8,000 (89,0%)	4,655

In experiment 5, the maximum ethanol production occurred after 12 hours (11.570 g/L). After this time, there was a reduction in the concentration of ethanol in the fermented medium and an increase in the concentration of acetic acid. The assumption that ethanol is converted into acetic acid is plausible, as the acetic bacteria produce 1.3 g of acetic acid stoichiometrically for every g of ethanol consumed. In other words, the values in brackets in Table 11 show the percentage conversion of ethanol to acetic acid.

6.5 Statistical Study of Experimental Fermentation Planning

The theoretical yield of substrate to product ($Y_{p/s}$), theoretical conversion (C) and Productivity (P) were calculated for the 8 and 12 hour times, which represented the highest ethanol yields (peak), and these values are shown in Table 10.

Table 10 Yp/S, P and C values at 8 and 12 hours of fermentation

Experi ment	Yeast	(NH$_4$)SO$_4$	KH$_2$PO$_4$	Yp/s (8 h)	C$_{(8h)}$	P$_{(8h)}$	Yp/s $_{(12)}$	C$_{(12)}$	P$_{(12)}$
1	5	0,3	0,05	0,267	0,523	0,963	0,311	0,609	0,748
2	15	0,3	0,05	0,347	0,681	1,253	0,338	0,662	0,813
3	5	0,9	0,05	0,328	0,643	1,185	0,349	0,684	0,841
4	15	0,9	0,05	0,386	0,756	1,393	0,361	0,707	0,868
5	5	0,3	0,19	0,376	0,737	1,358	0,401	0,785	0,964
6	15	0,3	0,19	0,371	0,726	1,337	0,383	0,751	0,921
7	5	0,9	0,19	0,415	0,813	1,497	0,395	0,773	0,949
8	15	0,9	0,19	0,454	0,888	1,636	0,374	0,732	0,898
9	10	0,6	0,12	0,461	0,901	1,661	0,374	0,732	0,898
10	10	0,6	0,12	0,382	0,748	1,378	0,322	0,629	0,774
11	10	0,6	0,12	0,354	0,692	1,275	0,0462	0,091	0,111

*Note: the units of Yp/s and C are expressed in g of ethanol/gram of substrate and that of P g/L.h

It can be seen from Table 10 that the best values were in experiments 8 and 9 with an 8-hour process, where the yields were 1.635 and 1.660 g/L.h, respectively, and the same is true for Yp/s and C. However, experiment 8 used more reagent than 9, making it more interesting to work under the conditions of the eighth test.

The data from the planning matrix was analyzed using *Statistica* version 5.0. Figures 8 and 9 show the Pareto graphs of the standardized effects of the Phosphorus and Ammonium Source concentrations and Yeast Concentration for $Y_{p/s}$, P and C.

Figure 8Pareto diagrams of yeast concentration effects Phosphorus and Ammonium source concentration in (A) Yp/S, C (B) and P (C) after eight hours of fermentation.

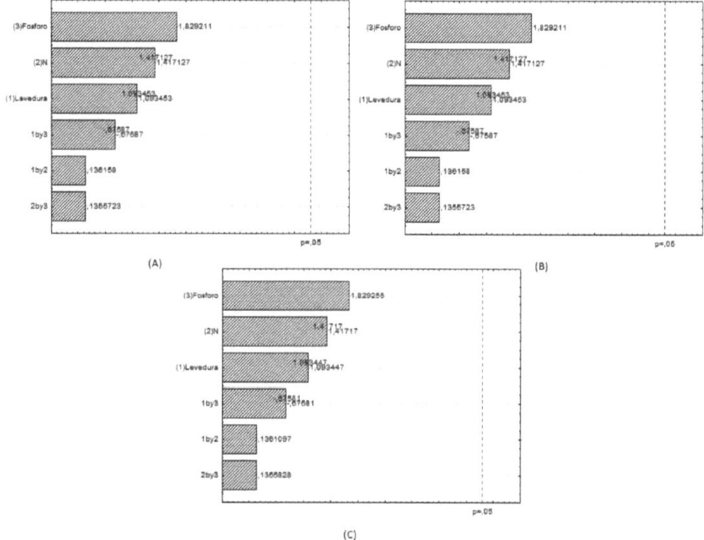

Source: Own authorship

Figure 9Pareto diagrams of yeast concentration effects Phosphorus and Ammonium source concentration in (A) Yp/S, C (B) and P (C) after twelve hours of fermentation.

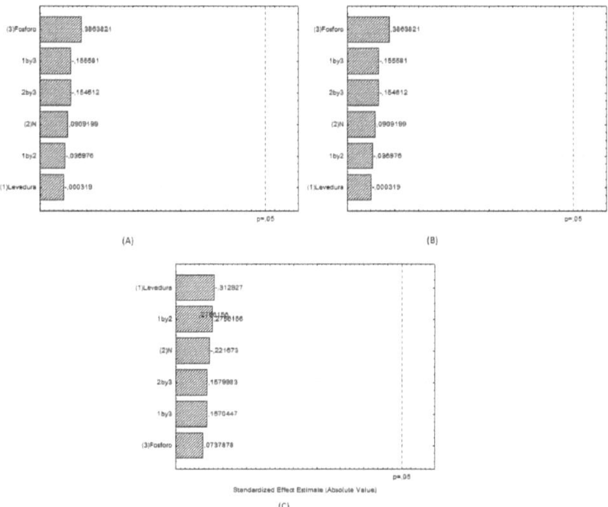

Source: Own authorship

It can be seen that all the effects (linear, quadratic and interaction) were not significant, considering the 95% confidence level (p-value <0.05). In other words, there were no influences at the levels studied for the input variables, only the averages were statistically significant. What can be said is that, in this study, the levels of the input variables did not influence the alcoholic fermentation process. Thus, the factorial design can be analyzed individually, and it can be seen that experiments 8 and 9 have the highest values for the kinetic parameters evaluated.

6.6 Enzyme kinetics

Enzymatic hydrolysis was carried out on the pre-treated bagasse, taking samples at different times, and then the RA was analyzed and quantified. These values are shown in Table 11.

Table 11 Sugar concentration at different times for different substrate concentrations

Time(min)	Substrate concentration							
	0.10%(g/L)	0.50%(g/L)	1%(g/L)	1.20%(g/L)	1.80%(g/L)	2%(g/L)	3%(g/L)	4.50%(g/L)
0	0,387185	0,387185	0,387185	0,387185	0,387185	0,387185	0,387185	0,387185
1	0,782933	1,199478	2,753112	2,894407	4,77834	5,98944	6,21698	6,766257
3	0,734612	1,165837	2,496823	2,537193	3,683457	5,823067	5,492767	6,262243
5	0,675283	1,278383	2,685217	2,682158	6,744237	4,317143	5,650577	7,322873
10	0,623288	1,359735	2,485813	2,61671	3,287097	4,856633	5,09763	6,81519
30	0,598822	1,260033	1,956722	2,836298	3,82781	5,84998	5,305597	6,70876
45	0,677727	1,378085	2,04419	3,205133	4,40767	4,40033	5,129437	7,998153
60	0,748068	1,47167	1,68453	3,20758	4,107953	3,855947	4,98386	8,204897
120	0,748068	1,250247	2,965972	2,907863	4,412563	4,98386	5,300703	8,486263

To better visualize this data, a graph of the concentration of RA as a function of time was plotted, as shown in Figure 10.

Figure 10Graph of concentration variation over time at different substrate concentrations

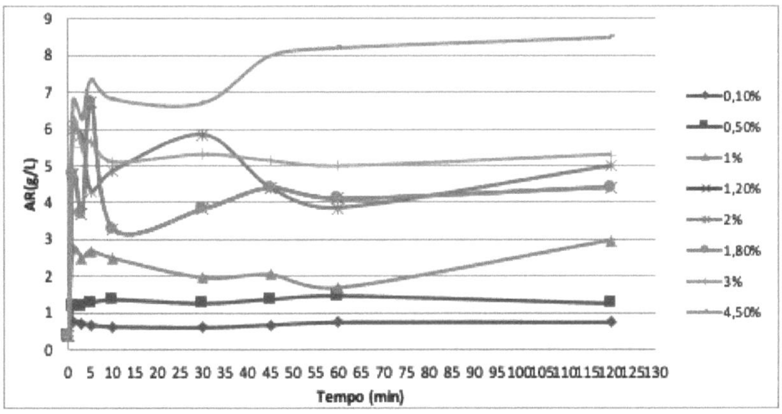

Source: Own authorship

It can be seen in Figure 10 that hydrolysis is very fast, with almost all the sugar being released in just one minute, after which production decreases, in some cases after which more sugar is produced or production becomes constant. There may have been inhibition of the enzyme by the product, adsorption of the enzyme with hemicellulose or lignin, clogging of the pores, desorption may not have taken place or even denaturation or loss of enzyme activity.

With these values, the initial velocities for each case were calculated in order to adjust these values with the Michalis-Menten equation. These values are shown in Table 12.

Table 12Initial velocity values

S_0 (g/L.s)	v_0 (g/L)
0,1	0,40
0,5	0,81
1	2,37
1,2	2,51
1,8	4,39
2	5,60
3	5,83
4,5	6,38

Using the data in Table 12, a graph was plotted (Figure 11) of the initial reaction speed as a function of the initial substrate concentration.

Figure 11Effect of total substrate (cellulose) on initial velocity

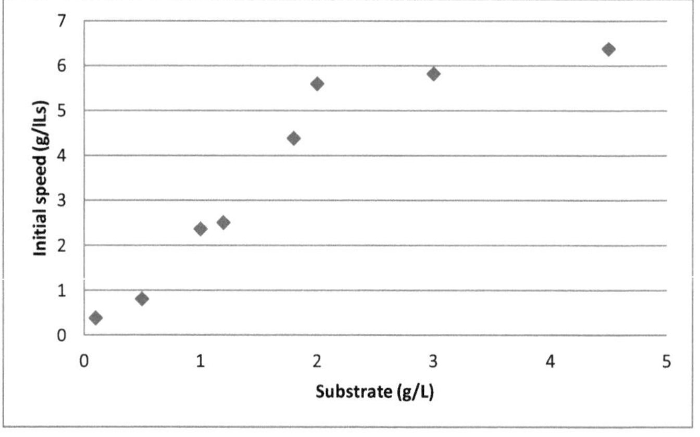

Source: Own authorship

The cellulose value was corrected for the initial substrate concentration in order to observe the change in the graph shown in Figure 11. The corrected data is shown in Table 13. A graph was plotted with this data, which is shown in Figure 12. The data was corrected by removing the influence of humidity on the concentration of cellulose in the substrate.

Table 13Initial velocity values with substrate concentration as cellulose.

S_0 (g/L.s)	v_0 (g/L)
0,03	0,40
0,16	0,81
0,32	2,37
0,39	2,51
0,58	4,39
0,65	5,60
0,97	5,83
1,46	6,38

Figure 12Effect of substrate (cellulose) on initial velocity

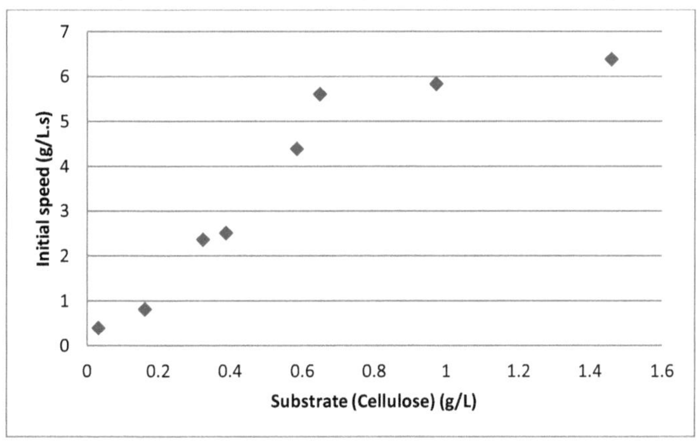

Source: Own authorship

Comparing Figures 11 and 12, it can be seen that these graphs show similar behavior, but the graph in Figure 11 is less steep than the other, thus showing greater accessibility of the enzyme.

In order to calculate the Michalis-Menten constants, the Lineweaver-Burk method was used, which consists of adjusting the Michaelis-Menten equation (Equation 1) by inverting both sides of it, obtaining Equation (14).

$$\frac{1}{V_0} = \frac{k_m}{V_m}\frac{1}{[S]} + \frac{1}{V_m} \quad (14)$$

Plotting the term $1/V_0$ against $1/[S]$ the angular coefficient will be the term k_m/V_m and the linear coefficient will be $1/V_m$. Thus, the graph (figure below) was made for both the complete substrate and cellulose alone.

Figure 13Lineweaver-Burk Methodology for Complete Substrate

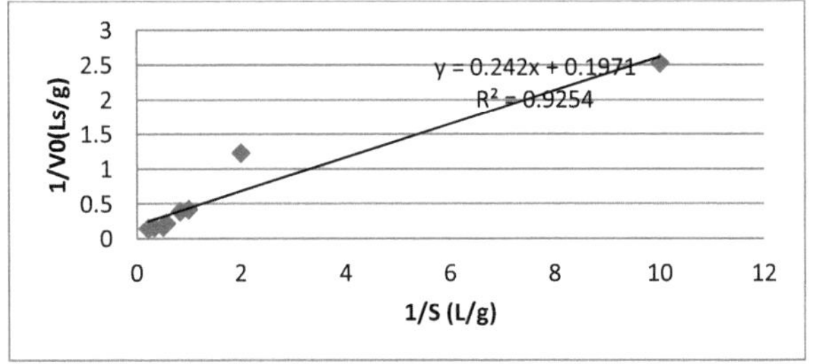

Source: Own authorship

Figure 14Lineweaver-Burk Methodology for Substrate Containing Only Cellulose

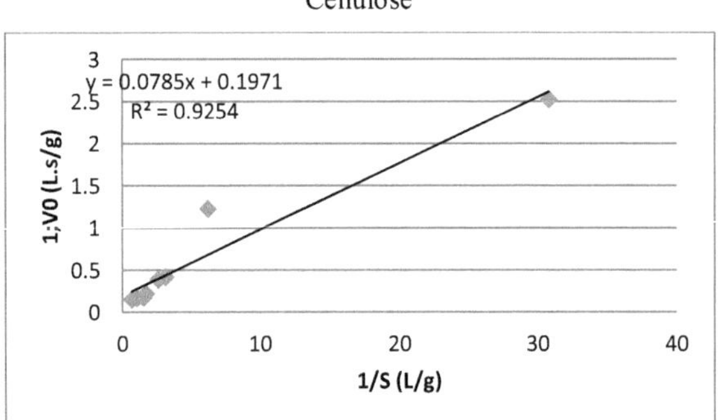

Source: Own authorship

Figures 13 and 14 show that for the complete substrate $V_m = 5.08$ and $K_m = 1.23$, while for the substrate containing only cellulose $V_m = 5.08$ and $K_m = 0.40$. Figure 19 shows greater accessibility for the substrate containing only cellulose than the complete substrate, which is confirmed by the K_m values, which were lower in the latter case.

6.7 Designing an Ideal CSTR

The initial substrate concentration was set at 10 g/L and the volumetric flow rate at 5 L/min (data for a bench project), with these values the conversion value was varied and with Equation (13) the reactor volume was calculated, so the result is shown in Figure 15.

Figure 15Reactor volume as a function of conversion

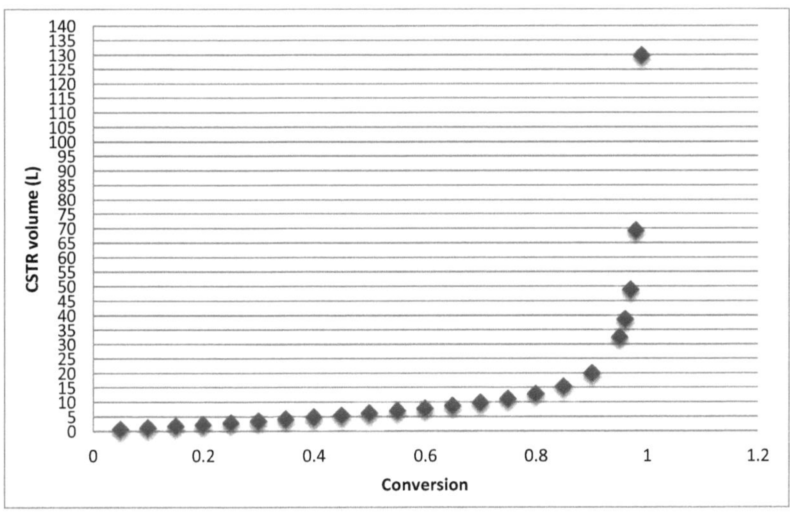

Source: Own authorship

Figure 23 shows a tendency for the reactor volume to increase exponentially as the conversion value increases. Thus, for the manufacture of a bench-scale reactor, a conversion of 90% can be chosen, which means making a reactor of approximately twenty liters.

7. CONCLUSIONS

Analyzing the hydrolysis results shows that the best pre-treatment was with sulfuric acid. The best experiment was the initial one (15 min, 0.5% acid, 0.05 FPU). However, it is advisable to use 7 FPU instead of 0.05 FPU of cellulase.

The best fermentation results were obtained in experiments 8 (15g/L yeast, 0.9 g/L ammonium and 0.19 phosphorus) and 9 (10g/L yeast, 0.6g/L ammonium and 0.12 phosphorus), which produced 13 g/L of ethanol, but it is advisable to use the parameters of experiment 9, as less reagent is used under these conditions. It is also clear that all the sugar is consumed in 8 hours and practically all the ethanol is produced in 8 hours of processing, so it would be more interesting to work with the concentrations from experiment 9 and stop the procedure after 8 hours of fermentation.

The statistical analysis of the fermentation experimental design showed that the effects were not significant to the response, considering a 95% reliability level. For future work, we suggest analyzing this design with the optimum values (peaks), rather than by time, and if this doesn't work, we recommend expanding the factorial design to a star configuration.

The enzymatic hydrolysis of cashew bagasse is very fast, and after a short time the sugar production decreases, or in some cases becomes constant. This can be explained by inhibition of the enzyme by the product, adsorption of the enzyme with hemicellulose or lignin, obstruction of the pores, desorption may not have taken place or even denaturation or loss of enzymatic activity.

It is also concluded that the Michaelis-Menten equation fits the values of this kinetics in 92% of the points. The design of the CSTR shows an almost exponential increase in volume as a function of the increase in conversion. To build such a reactor on a bench scale for a 90% conversion, we would have to use a reactor of approximately 20 liters. Further work could adjust these points with different models or even design different reactors.

8. REFERENCES

Abujamra, L. B.; Production of Alcoholic Distillate from Fermented Sweet Potato Mash, State University of São Paulo, 2009 (Doctoral Thesis).

ADENE, AREAC. Renewable Energies in Urban Environments, Portugal, 2005.

Amorim, B. C., Guimarães, D. P., Sousa, C. A. B Oliveira, I. N., e Oliveira, L. S. C. Avaliação das Características Físico-Químicas Do Resíduo Seco Do Caju Lavado E Sem Lavar Para Utilização Na Produção De Celulase. Anais XVII Congresso Brasileiro De Engenharia Química, Foz Do Iguaçu - PR, 2010.

Banjeree, G. Car, S. Liu, T. Williams, D. L. Meza, S. L. Walton J. D. Hodge, D. B. Scale-Up and Integration of Alkaline Hydrogen Peroxide Pretreatment, Enzymatic Hydrolysis, and Ethanolic Fermentation. Biotechnology and Bioengineering. Vol. 109, n. 4, p. 922 - 931. April 2012.

Beguin, P. Molecular biology of cellulose degradation. Annual Review of Microbiology, v. 44, p. 219 - 248, 1990.

Bobleter, O. Hydrothermal degradation of polymers derived from plants. Prog. Polym. Sci. v.19, pp. 797-841, 1994.

BRASIL. Métodos Físico-Químicos para Análises de Alimentos/Ministério da Saúde, Instituto Adolfo Lutz, 2005.

Caldas, R. C. R. B.,. Soares, I. B., Benachour M., Abreu, C. A. M. Study of the Acid Hydrolysis Process of Sugarcane Bagasse Using Lewis Acids. Anais XVII Congresso Brasileiro De Engenharia Química, Foz Do Iguaçu - PR, 2010.

Carminal, G.; Santin, J. L.; Sola, C. Kinetic modeling of the enzymatic hydrolysis of pretreated cellulose. Biotechnol. Bioeng., v. 27, p. 1282-1290, 1985.

Carraro F.;Cunha, M. M. Manual de Exportação De Frutas. Brasília: Marra- Sdr-Frupex/Iilca, 1994. P. 254.

Carvalho, G. B.M., Ginoris, Y. P., Cândido, E. J., Canilha, L., Carvalho, W., Silva, J. B. A. Study of Eucalyptus Hydrolysate in Different Concentrations Using Vacuum Evaporation for Fermentation Purposes. Revista Analytica v 14, pag 54-57, 2005.

Cassales, A. R. ; Optimization of the Hydrolysis of Soybean Hulls (*Glycine max*) and Evaluation of the Ability of Microorganisms to Produce Xylitol and Ethanol from this Hydrolysate. 2010. Dissertation (Master's Degree in Food Science and Technology). Federal University of Rio Grande do Sul.

Corazza, M. L.; Rodrigues,D. G.; Nozaki, J.; Quím. Nova São Paulo Jul/Aug. 2001, vol.24 no.4.

EMBRAPA. Technologies. 2004. Available (at: www.cnpgc.embrapa.br/tecnologias/quersabermais/500p/P256.html). Accessed on: January 23, 2004.

Brazilian Agricultural Research Corporation. The world of the cajulino. Fortaleza: EMBRAPA/CNPAT. 1999 20 p.

Faria, F.S.E.D.V. Influence of two Saccharomyces cerevisiae strains on the production of cashew fermented products (Anacardium occidentale, L.) under different fermentation conditions. 1994. 99f. Dissertation (Master's in Food Technology). Federal University of Ceará, Fortaleza.

Garcia, D. R. Determination of Kinetic Data for the Pretreatment of Sugarcane Bagasse with Alkaline Hydrogen Peroxide and Subsequent Enzymatic Hydrolysis. 2009. Dissertation (Master's Degree in Chemical Engineering). State University of Campinas.

Garruti, D.S. Composition of volatiles and aroma quality of cashew wine. Campinas-SP. State University of Campinas (Doctoral thesis), 2001.

Garruti, D.S.; Cassimiro, A.R.S.; Abreu, F.A.P. Agro-industrial process: making cashew ferment. Technical communication 82. Fortaleza, CE - October, 2003.

Gilkes, N. R.; Langsford, M. L.; Kilburn, D. G.; Miller, R. C. Jr.; Warrem, R. A. J. Mode of action and substrate specifities of cellulases from cloned bacterial genes. Journal of Biological Chemistry, v. 259, n. 16, p. 10455-10459, 1991.

GLOBO RURAL. Reform in the house, Globo Rural, Agropecuária, Negócios e Vida no Campo, March, no. 233, p. 58-63, 2005.

Guedes, R. C.. Rocha, C. E. M Silva, N. M. P. Cruz, P. P.. Santos E. S Macedo, G. R.. Study of Lignocellulosic Waste Pre-Treatment in the Induction of Cellulase Enzyme Synthesis. Anais XVIII Congresso Brasileiro de Engenharia Química, Foz do Iguaçu - PR, 2010.

Gutierrez, L.E.; Accumulation of Trehalose in *Saccharomyces* Strains During Alcoholic Fermentation, Anais ESALQ, Piracicaba, vol 47, p 597-608, 1990.

Gutierrez, L.E; Glycerol Production by *Saccharomyces* Strains During Alcoholic Fermentation; Anais ESALQ, Piracicaba,vol 48, p 55-69, 1991.

Hasif, R. Jason, W.S.Y. Cherrington, T. Danquah, M. K. Exploring alkaline pretreatment of microalgal biomass for bioethanol production. Applied Energy, v.88 p. 3464-3467, 2011.

Holanda, J. S.; Oliveira, A. C. Protein enrichment of cashew stalks using yeasts for animal feed. Pesquisa Agropecuária Brasileira, v. 33, n. 5, p. 787-792, 2001.

Knauf, M.; Moniruzzaman, M. Lignocellulosic biomass processing: a perspective. International sugar journal. 106(1263):147-150, 2004.

Koo, B.W. Kim H. Y. Park, N. Lee, S. M. Yeo, H. Choi I. G. Organosolv pretreatment of Liriodendron tulipifera and simultaneous saccharification and fermentation for bioethanol production. Biomass and Bioengergy, v. 35, p. 1833-1840, 2011.

Kuhad, K.C., Sing, A. (1993), "Lignocellulose Biotechnology: Current and Future Prospects". Critical Reviews in Biotechnology, v.13, n2, p 151-172.

Lehninger, A.L. (1976), Biochemistry: molecular components of cells, v.1, p.182.

Leite, I. R., Faria, J. R.,. Carvalho, L. F., Ribeiro, E. J. Cardoso, V. L. Avaliação Da Ação De Antibiótico Natural Na Fermentação Alcoólica Contaminada Por Cultura Mista. Anais XVII Congresso Brasileiro De Engenharia Química, Foz Do Iguaçu - PR, 2010.

Leite, L.A.S. The cashew agroindustry in Brazil. Public policies and economic transformations. Fortaleza: EMBRAPA - CNPAT, 1994. 195 p.

Mariotto, J. R; Enzymatic Kinetics Workbook. 2006. Available at http://www.enq.ufsc.br/labs/probio/disc_eng_bioq/apostilas/Apostila_cinetica_enzimatica_ju.pdf. Accessed on 24/08/2012.

Martín, C.; Gonzáles, Y.; Fernández, T.; Thomsen, A.B.; Investigation of cellulose convertibility and ethanolic fermentation of sugarcane bagasse pretreated by wet oxidation and steam explosion. Journal of Chemical Technology and Biotechnology, Oxford, v.81, p.1669-1677, 2006.

Martínez, M. J. Fungal pretreatment: An alternative in second-generation ethanol from wheat straw. Bioresource Technology, v.102, p.7500-7506.

Maziero P., Gonçalves A.R, Polikarpov I. Rocha G.J.M. Estudo Cinético Do Pré-Tratamento Hidrotérmico Para Obtenção De Etanol De Segunda Geração. Anais XVIII Congresso Brasileiro de Engenharia Química, Foz do Iguaçu - PR, 2010.

MENEZES, C. R.; SILVA, I. S.; DURRANT, L. R. Sugarcane bagasse: a source for the production of ligninocellulolytic enzymes. Estudos Tecnológicos, v. 5, n. 1, p. 68-78, 2009.

Menezes, J.B.; Alves, R.E.; Physiology and Post-harvest Technology of the Cashew Stalk. Fortaleza: EMBRAPA-CNPAT, 1995. P.20. (EMBRAPA, CNPAT, Documentos, 17). ISSN 0103-5797.

Milagres, A. M. F., Borges, L., Aguiar, C. L. Degumming of ramie for textile purposes using enzymatic extracts. Anais do SHEB, n. 4, p. 261269,1994.

MILLER, G. L. Use of dinitrosalicylic acid reagent for determination of reducing sugar. Analytical Chemistry, v. 31, n. 3, p. 426-428, 1959.

Moreira Neto, J. Mathematical Modeling of the Enzymatic Hydrolysis Process of Sugarcane Bagasse Subjected to Different Pretreatments.2011. Dissertation (Master's Degree in Chemical Engineering). State University of Campinas. Campinas -SP.

Moreschi, S. R. M. Hydrolysis, with subcritical water and CO_2 , of the starch and cellulose present in the residue from the supercritical extraction of ginger (Zingiber officinale Roscoe): production of oligosaccharides. State University of Campinas. Campinas, SP, 2004 (Doctoral thesis).

Moutta, R. O.,. Rodrigues, R. C. L. B, Silva,S. S. Bioethanol Production From Hemicellulosic Hydrolysate Obtained From Sugarcane Straw. Anais XVII Congresso Brasileiro De Engenharia Química, Foz Do Iguaçu - PR, 2010.

Niedetzky, B.; Steiner, W. A new approach for modeling cellulase-cellulose adsorption and the kinetics of the enzymatic hydrolysis of microcrystalline substrate. Biotechnol. Bieng., v. 42, p. 469-479, 1993.

Nobre, T. P; Hori, J.; Alcarde, A. R; Ciênc. Tecnol. Aliment., Campinas, 27: jan.-mar. 2007, p 20-25, vol 27.

Oberoi, H. S. Vadlani, P. V. Madl, R. L. Saida, L. Abeykoon, J. P. Ethanol Production from Orange Peels: Two-Stage Hydrolysis and Fermentation Studies Using Optimized Parameters through Experimental Design. Journal of Agricultural and Food Chemistry. V. 58, p. 3422-3429.

Oliveira, F.M.V. Pinheiro, I.O. Gonçalves, A.R. Rocha, G.J.M.. Effect of Dilute Acid Pre-Treatment and Alkaline Delignification on the Enzymatic Hydrolysis of Sugarcane Straw to Obtain Ethanol. Anais XVIII Congresso Brasileiro de Engenharia Química, Foz do Iguaçu - PR, 2010.

Parente, J.I.G.; Pessoa, P.F.A.P.; Namekata, Y. Diretrizes para recuperação da cajucultura no Nordeste. Fortaleza: Embrapa. March, 1991.

Petinari, R.A.; Tarsitano, M.A.A. Commercialization of fresh cashew in the Northwest region of the state of São Paulo. Revista Brasileira de Fruticultura, V.24, n. 3, p. 700-702. Dec. 2002.

Pietrobon, V. Hydrolysis of sugarcane bagasse pretreated with acid and alkali using commercial microbial enzymes. University of São Paulo, Luiz Queiroz College of Agriculture. 2008. Piracicaba - SP.

Reyes, J., Peralta-Zamora, P. , Durán, N.; Enzymatic Hydrolysis of Rice Husk Using Cellulases. Effect of Chemical and Photochemical Treatments. Química Nova, v 21, p 140-143, 1998.

Rocha, A. S., Silva, F. L. H., Conrado, L. S., Lima, F. C. S., Lima, E. E., . Carvalho, J. P. D, Guimarães, D. P. Evaluation of Acid Pretreatment of Cashew Stalk Bagasse for the Production of Fermentable Sugars. Proceedings of the XVIII National Bioprocess Symposium, Caxias do Sul, July 2011.

Rocha, M. V. P., PRODUCTION OF BIOETHANOL FROM POME FRUIT (Anacardium occidentale L.) BY SUBMERSE FERMENTATION. 2010. Dissertation (Postgraduate course in Chemical Engineering). Federal University of Rio Grande do Norte.

Rodrigues T. H, Pinheiro, A. D. T., Rego, D. R. B., Rocha M. V. P.,. Gonçalves L. R. B., Macedo G. R. Evaluation of the Potential of Cashew Stem (Anacardium Occidentale L.) for Bioethanol Production by Saccharomyces Cerevisiae. Anais XVII Congresso Brasileiro De Engenharia Química, Foz Do Iguaçu - PR, 2010.

Rosillo-Calle F.; Cortez L. A. B. Towards ProAlcool II - A review of the Brazilian bioethanol program. Biomass Bioenergy, v. 14, p. 115-124, 1998.

Rueda. S. M. G, Santander C. G., Costa, A. C., Maciel Filho R. Effect of Pretreatment with Dilute Phosphoric Acid on the Enzymatic Hydrolysis of Sugarcane Bagasse for the Production of Fermentable Sugars. Anais XVIII Congresso Brasileiro de Engenharia Química, Foz do Iguaçu - PR, 2010.

Ruggiero C., 2002. Available at: www.todafruta.com.br. Accessed on: June 15, 2009 at 16:22.

Salvachúa, D.; Prieto, A.; López-Abelairas, M.; Lu-Chau T. ; Martínez, Á. T.; Fungal pretreatment: An alternative in second-generation ethanol from wheat straw. Bioresour Technoligic. 2011 Aug;102(16):7500-6. 2011.

Santos, J. R. A. Gouveia, E. R. PRODUÇÃO DE BIOETANOL DE BAGAÇO DE CANA-DE-AÇÚCAR Revista Brasileira de Produtos Agroindustriais, Campina Grande, v.11, n.1, p.27-33, 2009.

Santos, J. R., Pinheiro, I. O., Souto-Maior, A. M., Gouveia, E. R., Estudo Cinético da Hidrólise do Bagaço de Cana-de-Açúcar, Anais XVII Congresso Brasileiro de Engenharia Química, Recife, 2008 .

Sattler, W.; Esterbauer, H.; Glatter, O.; Steiner, W. The effect of enzyme concentration on the rate of hydrolysis of cellulose, Biotechnol. Bioeng., v. 33, p. 1221-1234, 1989.

Silva, A. B., Prado, A. G., Morais Junior M. A., Menezes, R. S. C. Evaluation of Forage Palm Hydrolysate for Bioethanol Production. Anais XVII Congresso Brasileiro De Engenharia Química, Foz Do Iguaçu - PR, 2010.

Silva, M. E., Araújo G. T., Alves J. J. N. Kinetic Evaluation of Cashew Stem Fermentate (Anacardium Occidentale L.) Aimed at Obtaining Hydrated Ethanol. Anais XVII Congresso Brasileiro De Engenharia Química, Foz Do Iguaçu - PR, 2010.

Soares, I. B., Bois, B. J., Baudel, H. M., Benachour, M., Abreu, C. A. M., Effect of Washing, Grinding and Cellulase Loading on the Enzymatic Hydrolysis of Sugarcane Bagasse Pretreated by Steam Explosion, Anais XVII Congresso Brasileiro de Engenharia Química, Recife - PE, 2008.

Sousa, M., Mesquita, F.M.R.;. Freitas, M.M.M , Santos J.C.S.;Sá, T.N.M, Pinto,G.A.S. Fermentação Alcoólica Utilizando Juco Do Pedúnculo De Caju (Anacardium Occidentale L.). Anais XVII Congresso Brasileiro De Engenharia Química, Foz Do Iguaçu - PR, 2010.

TAVARES, M. B. R. Study of the production of the enzyme cellulase from the lignocellulosic material cashew bagasse by semi-solid fermentation using Aspergillus niger. 2009. 108f. Dissertation (Master's in Chemical Engineering), Federal University of Campina Grande, Campina Grande, PB.

Tsao, G. T., Cellulosic material as a renewable resource. Process Biochemistry, v. 10, p. 12-14, 1978.

Tsao, G. T., Cellulosic material as a renewable resource. Process Biochemistry, v. 10, p. 12-14, 1978.

Updegraff, D. M. (1969) Semimicro determination of cellulose inbiological materials. Analytical Biochemistry, v.32, p.420-424, December. Yoshikuni, T. Noriko, T. Seung-Hwan, L. Takashi, E. Pretreatment of Eucalyptus Wood Chips for Enzymatic Saccharification Using Combined Sulfuric Acid-Free Ethanol Cooking and Ball Milling. Biotechnology and Bioengineering, Vol. 99, No. 1, p. 75-85, January 1, 2008.

V. F. N. Da Silva, F. M. V. Oliveira, A. R. Gonçalves and G. J. M. Rocha. Study of the Dilute Acid Pretreatment of Sugarcane Straw on a Pilot Scale as a Step in the Process of Obtaining Cellulosic Ethanol. Anais Xviii Congresso Brasileiro De Engenharia Química, Foz Do Iguaçu - Pr, 2010.

Vasconcelos1, S. M. Souza, G. K. M. Santos A. M. P. Soutomaior A. M. Estudo Da Hidróólise Enzimática Do Bagaço De Cana-De-Açúcar Submetido a Diferentes Condições De Pré-Tratamento Com Acid Posfórico. Anais XVII Congresso Brasileiro De Engenharia Química, Foz Do Iguaçu - PR, 2010.

Volp P. L. O.; Study of the Alcoholic Fermentation of Dilute Solutions of Different Sugars Using Flow Microcalorimetry; Química Nova, vol 20, p 528-534, 1997.

Zhao, X.; Peng, F.; Cheng, K.; Liu, D. Enhancement of the enzymatic digestibility of sugarcane bagasse by alkali-peracetic pretreatment. Enzyme Microb. Technol. v. 44, p. 17-23, 2009.

Appendix A - Glucose consumption and ethanol production tables for all the experiments as a function of time.

Table A1. Glucose consumption over time in the eleven experiments

Time	Exp 1	Exp 2	Exp 3	Exp 4	Exp 5	Exp 6	Exp 7	Exp 8	Exp 9	Exp 10	Exp 11
0	28,84268	28,84268	28,843	28,84268	28,84268	28,843	28,843	28,843	28,843	28,843	28,843
2	18,9242	2,77965	20,5426	17,78214	6,69967	6,69967	16,82352	15,61632	17,64013	6,25681	12,66715
6	0,92224	0,17552	0,86721	9,47791	1,04286	0,7295	0,65597	0,62212	0,72915	0,72915	0
8	0	0	0	0	0	0	0	0,84206	0	0,4873	0
12	1,52253	0,92724	0	0	0	0	0	0	0	0	0
24	0	0	0	0	0	0	0	0	0	0	0
36	1,41239	0	0	0	0	0	0	0	0	0	0
48	0	0	0	0	0	0	0	0	0	0	0
72	0	0	0	0	0	0	0	0	0	0	0

*Obs All data is given in g/L

Table A2 Ethanol production over time in the eleven experiments

Time	Exp 1	Exp 2	Exp 3	Exp 4	Exp 5	Exp 6	Exp 7	Exp 8	Exp 9	Exp 10	Exp 11
0	0	0	0	0	0	0	0	0	0	0	0
2	1,56091	3,25702	2,32429	2,98434	0,78244	0,78244	2,75913	3,91402	3,12951	2,52091	3,66493
6	5,97675	4,94824	9,61275	9,47791	12,31327	10,93751	12,16722	10,94831	7,7102	7,7102	10,05388
8	7,702334	10,02868	9,47791	11,14793	10,86029	10,69636	11,97835	13,08563	13,28426	11,02139	10,20021
12	8,98201	9,75928	10,0836	10,42147	11,57062	11,05678	11,38795	10,78207	10,78392	9,28466	1,33313
24	9,15215	8,62284	9,89155	9,16527	9,41591	8,30583	9,77764	8,44337	9,26474	8,74808	9,58448
36	6,8507	7,55693	8,2345	8,53603	8,30564	8,22792	6,40316	6,24754	0,51065	5,86542	6,58215
48	8,06089	6,74562	5,72033	6,21083	8,23033	6,30367	1,00907	3,71948	1,53906	4,2296	6,80012
72	2,11627	2,03228	6,02187	4,35748	4,65521	3,45426	2,46217	2,49015	3,67538	3,62936	0,42273

*Obs All data is given in g/L

I want morebooks!

Buy your books fast and straightforward online - at one of world's fastest growing online book stores! Environmentally sound due to Print-on-Demand technologies.

Buy your books online at
www.morebooks.shop

Kaufen Sie Ihre Bücher schnell und unkompliziert online – auf einer der am schnellsten wachsenden Buchhandelsplattformen weltweit! Dank Print-On-Demand umwelt- und ressourcenschonend produziert.

Bücher schneller online kaufen
www.morebooks.shop

info@omniscriptum.com
www.omniscriptum.com

Printed by Books on Demand GmbH, Norderstedt / Germany